Problem Solvers

Edited by L. Marder
Senior Lecturer in Mathematics, University of Southampton

No. 1

Ordinary Differential Equations

JOHN HEADING, M.A., Ph.D.

Professor of Applied Mathematics,
University College of Wales, Aberystwyth

LONDON · GEORGE ALLEN & UNWIN LTD
RUSKIN HOUSE   MUSEUM STREET

First published in 1971

ISBN 0 04 517002 9 *paper*
    0 04 517003 7 *cased*

Printed in Great Britain
in 10 on 12 pt 'Monophoto' Times Mathematics Series 569
by Page Bros. (Norwich) Ltd., Norwich

Preface

Whereas *practical problems* in science and engineering require the ultimate *numerical* solution of differential equations, yet the study of scientific and engineering *principles* requires the exact or approximate *analytical* solution of such equations. The object of this book is to concentrate solely upon the latter aspect of the subject, by providing solutions for as many different types of soluble differential equations as possible. The underlying theory has been touched upon only in a few brief remarks. The approach is elementary, in that what is covered refers mainly to first- and second-year degree work, though both simple and advanced examples are included in varying degrees. Out of the vast range of topics that might have been included in the two concluding chapters, a choice had to be made depending on the limited space available. The choice of the Laplace transform and Bessel functions is, to some extent, arbitrary, but no one who uses differential equations, or who reads widely, can remain in ignorance of these topics for long. They certainly form a suitable introduction to further transform methods and to more advanced transcendental equations.

The notation for derivatives, dy/dx, $d^2y/dx^2, \ldots$, printed in the form of a fraction, is notorious for consuming page space. In order to avoid this waste of space, the simpler notation Dy, $D^2y, \ldots$, has been used throughout, although in some places the more standard notation has had to be used to avoid confusion.

I must thank my one-time colleague Dr. L. Marder for suggesting that I should write this book for the series of which he is the General Editor.

J. HEADING

Contents

Chapter 1
The Formation of Differential Equations

If y is given as an arbitrary linear combination of n distinct functions of x,
$$y = A_1 f_1(x) + A_2 f_2(x) + \ldots + A_n f_n(x),$$
a differential equation of order n is obtained by eliminating the n constants. Evaluate $Dy, D^2y \ldots, D^n y$, and the constants are eliminated usually by means of a determinant. It is necessary for the coefficient of $D^n y$ (known as the Wronskian of the n functions $f_i(x)$) not to vanish; the n functions are then independent.

If, in addition,
$$z = A_1 g_1(x) + A_2 g_2(x) + \ldots + A_n g_n(x),$$

denotes a second linear combination with the same constants, and if $D^n y$ and $D^n z$ are avoided, then elimination yields two simultaneous differential equations (obviously with a wide variety of dependent possibilities).

Differential equations may arise from the defining geometrical properties of a family of curves in a plane. More usually, equations (often simultaneous) arise from the application of physical laws to particular problems, since the physical laws refer to the time rate of change or the space rate of change of the physical variables involved.

Problem 1.1 Without using determinants, find the second order differential equation satisfied by the functions e^x, e^{2x}.

Solution. We eliminate A and B from
$$y = Ae^x + Be^{2x} \tag{1}$$
by forming the derivatives
$$Dy = Ae^x + 2Be^{2x}, \tag{2}$$
$$D^2 y = Ae^x + 4Be^{2x}. \tag{3}$$
An obvious procedure for eliminating A and B is to subtract (2) from (1) and (3) from (2):
$$Dy - y = Be^{2x},$$
$$D^2 y - Dy = 2Be^{2x}.$$
The elimination of B yields
$$D^2 y - 3Dy + 2y = 0. \qquad \square$$

Problem 1.2 Find the differential equation of the third order possessing the three independent solutions x, x^2, x^3.

  Solution. Let
$$y = Ax + Bx^2 + Cx^3,$$
with
$$Dy = A + 2Bx + 3Cx^2,$$
$$D^2y = \qquad 2B + 6Cx,$$
$$D^3y = \qquad\qquad 6C.$$

Elimination of A, B and C by means of a determinant yields
$$\begin{vmatrix} y & x & x^2 & x^3 \\ Dy & 1 & 2x & 3x^2 \\ D^2y & 0 & 2 & 6x \\ D^3y & 0 & 0 & 6 \end{vmatrix} = 0,$$

simplifying to
$$x^3 D^3y - 3x^2 D^2y + 6x\, Dy - 6y = 0. \qquad\qquad \square$$

Problem 1.3 Two simultaneous differential equations possess the two pairs of solutions
$$u = e^x\cos x, \qquad v = -\sin x;$$
$$u = e^x\sin x, \qquad v = \cos x.$$

Find two equations possessing these solutions.

  Solution. Let
$$u = Ae^x\cos x + Be^x\sin x,$$
$$v = -A\sin x + B\cos x.$$

Differentiation yields
$$Du = A(e^x\cos x - e^x\sin x) + B(e^x\sin x + e^x\cos x),$$
$$Dv = -A\cos x - B\sin x.$$

We eliminate A and B firstly from u, v, Du, and secondly from u, v, Dv:

$$\begin{vmatrix} u & e^x\cos x & e^x\sin x \\ v & -\sin x & \cos x \\ Du & e^x(\cos x - \sin x) & e^x(\sin x + \cos x) \end{vmatrix} = 0;$$

$$\begin{vmatrix} u & e^x\cos x & e^x\sin x \\ v & -\sin x & \cos x \\ Dv & -\cos x & -\sin x \end{vmatrix} = 0.$$

These simplify to
$$Du = e^x v + u, \qquad Dv = -e^{-x}u. \qquad\qquad \square$$

2

Problem 1.4 In Cartesian coordinates, find the differential equation representing all circles that may be drawn in the x,y-plane.

Solution. The general circle in the plane is represented by

$$x^2 + y^2 + 2fx + 2gy + h = 0,$$

with centre $(-f, -g)$ and radius $\sqrt{(f^2 + g^2 - h)}$.

Three arbitrary constants are involved, so we must differentiate three times with respect to x. After division by two, we obtain

$$x + y\,Dy + f + g\,Dy = 0,$$
$$1 + (Dy)^2 + y\,D^2y + g\,D^2y = 0,$$
$$3Dy\,D^2y + y\,D^3y + g\,D^3y = 0.$$

Elimination of the constants f, g and h yields

$$\begin{vmatrix} x^2 + y^2 & 2x & 2y & 1 \\ x + y\,Dy & 1 & Dy & 0 \\ 1 + (Dy)^2 + y\,D^2y & 0 & D^2y & 0 \\ 2Dy\,D^2y + y\,D^3y & 0 & D^3y & 0 \end{vmatrix} = 0.$$

Evidently,

$$\begin{vmatrix} 1 + (Dy)^2 + y\,D^2y & D^2y \\ 2Dy\,D^2y + y\,D^3y & D^3y \end{vmatrix} = 0,$$

simplifying to

$$D^3y[1 + (Dy)^2] - 3Dy(D^2y)^2 = 0. \qquad \square$$

Problem 1.5 Three tanks U, V and W, each of uniform horizontal cross-sectional area A, contain liquid at depths u, v and w respectively. Liquid flows from U to V at a rate equal to αA times the depth of liquid in U, and liquid flows from V to W at a rate equal to βA times the height of liquid in V. Find the differential equation satisfied by the height w.

Solution. For tank U, liquid is leaving only; the rate of increase of the volume Au is given to equal $-(\alpha A)u$; hence

$$Du = -\alpha u, \tag{1}$$

where $D \equiv d/dt$. For tank V, liquid enters and liquid leaves; the total rate of increase of volume Bv (and hence of v) is given by

$$Dv = \alpha u - \beta v. \tag{2}$$

For tank W, liquid enters only, and it is given that

$$Dw = \beta v. \tag{3}$$

We eliminate u between (1) and (2). (2) rearranged and differentiated gives Du:

$$D^2v + \beta\,Dv = \alpha\,Du.$$

3

(1) and (2) added give

$$D(u+v) = -\beta v,$$
$$Du = -Dv - \beta v.$$

Hence elimination of Du yields

$$D^2v + \beta\, Dv = \alpha(-Dv - \beta v),$$

or

$$D^2v + (\alpha + \beta)\, Dv + \alpha\beta v = 0.$$

Differentiating this, and noting from (3) that $Dv \propto D^2w$, we finally have

$$D^3w + (\alpha + \beta)\, D^2w + \alpha\beta\, Dw = 0.$$

More systematically, equations (1), (2) and its derivative, (3) and its two derivatives:

$$\alpha u + Du = 0,$$
$$\alpha u - \beta v - Dv = 0, \quad \alpha\, Du - \beta\, Dv - D^2v = 0,$$
$$\beta v - Dw = 0, \quad \beta\, Dv - D^2w = 0, \quad \beta\, D^2v - D^3w = 0$$

represent six equations in the quantities u, Du, v, Dv, D^2v and terms in w. A determinant yields

$$\begin{vmatrix} \alpha & 1 & 0 & 0 & 0 & 0 \\ \alpha & 0 & -\beta & -1 & 0 & 0 \\ 0 & \alpha & 0 & -\beta & -1 & 0 \\ 0 & 0 & \beta & 0 & 0 & -Dw \\ 0 & 0 & 0 & \beta & 0 & -D^2w \\ 0 & 0 & 0 & 0 & \beta & -D^3w \end{vmatrix} = 0,$$

reducing to the previously established equation. $\qquad\square$

Problem 1.6 When the separated function $\phi = X(x)\,Y(y)$ is substituted into the partial differential equation

$$\frac{\partial^2\phi}{\partial x^2} + \frac{\partial\phi}{\partial x} + \frac{\partial^2\phi}{\partial y^2} = 0,$$

find the ordinary differential equations satisfied by X and Y.

Solution. Writing $D_x \equiv d/dx$, $D_y \equiv d/dy$, we note that

$$\frac{\partial\phi}{\partial x} = (D_x X)Y, \qquad \frac{\partial^2\phi}{\partial x^2} = (D_x^2 X)Y,$$

and so on. Substitution into the equation yields

$$(D_x^2 X)Y + (D_x X)Y + X D_y^2 Y = 0.$$

We divide by XY to separate out completely terms in x and y:

$$\frac{D_x^2 + D_x X}{X} = -\frac{D_y^2 Y}{Y}.$$

The left-hand side is a function of x alone; the right-hand side is a function of y alone. Both sides must equal a constant, n^2 say. Then

$$D_x^2 + D_x X - n^2 X = 0,$$
$$D_y^2 Y + n^2 Y = 0.$$

The original partial differential equation thus yields *different* equations for X (and Y) for *different* values of n^2.

Problem 1.7 Find the ordinary differential equations that yield the separated solutions of the partial differential equation

$$\frac{\partial^2 \phi}{\partial r^2} + \frac{2}{r}\frac{\partial \phi}{\partial r} + \frac{1}{r^2}\frac{\partial^2 \phi}{\partial \theta^2} + \frac{\cot \theta}{r^2}\frac{\partial \phi}{\partial \theta} = 0.$$

Solution. Separated solutions, when possible, have the form

$$\phi = R(r)\,\Theta(\theta),$$

where R is a function of r alone, and Θ of θ alone. Substitution yields

$$\frac{d^2 R}{dr^2} + \frac{2}{r}\frac{dR}{dr} + \frac{1}{r^2}R\frac{d^2\Theta}{d\theta^2} + \frac{\cot\theta}{r^2}R\frac{d\Theta}{d\theta} = 0.$$

Division by $R\Theta/r^2$ enables us to separate the r and θ terms thus:

$$\left[\frac{d^2 R}{dr^2} + \frac{2}{r}\frac{dR}{dr}\right]\Big/\frac{R}{r^2} = -\left[\frac{d^2\Theta}{d\theta^2} + \cot\theta\frac{d\Theta}{d\theta}\right]\Big/\Theta.$$

The left-hand side is a function of r only, while the right-hand side is a function of θ only. Both sides must therefore equal a constant, k say. Subsequent development shows that the best form for this constant is $n(n+1)$, yielding the two ordinary differential equations

$$\frac{d^2 R}{dr^2} + \frac{2}{r}\frac{dR}{dr} - n(n+1)\frac{R}{r^2} = 0,$$

$$\frac{d^2\Theta}{d\theta^2} + \cot\theta\frac{d\Theta}{d\theta} + n(n+1)\Theta = 0.$$

It may be verified by substitution that

$$R = Ar^n + Br^{-n-1} \qquad (n \neq -\tfrac{1}{2}),$$
$$R = (A + B\log r)r^{-\frac{1}{2}} \qquad (n = -\tfrac{1}{2}),$$

and that simple solutions for Θ (such that $\Theta = 1$ when $\theta = 0$) are

$$\Theta = \cos \theta \qquad (n = 1),$$
$$\Theta = \tfrac{1}{2}(3 \cos^2 \theta - 1) \qquad (n = 2).$$

Hence two simple separated solutions are given by

$$\phi = (Ar + Br^{-2}) \cos \theta,$$
$$\phi = (Ar^2 + Br^{-3}) . \tfrac{1}{2}(3 \cos^2 \theta - 1).$$

Note. To be finite when $r = 0$, we must choose $B \equiv 0$. To remain finite as $r \to \infty$, we must choose $A \equiv 0$.

EXERCISES

1. Find the second order differential equation possessing the two solutions

$$y_1 = x^{-\frac{1}{2}} \sin x, \qquad y_2 = x^{-\frac{1}{2}} \cos x.$$

2. Find two simultaneous differential equations (the first not containing Dv and the second not containing Du) satisfied by the two pairs of solutions

$$u = e^x, \quad v = -e^x; \qquad u = e^{-5x}, \quad v = e^{-5x}.$$

3. A point particle P is attracted to a centre of force O by the radial force $f(r)$ per unit mass, where r is the distance OP. Using the acceleration components in polar coordinates in a plane, obtain the differential equation satisfied by the dependent variable $u \equiv 1/r$ in terms of the independent variable θ, where h is the given angular momentum of the particle.

4. Find the equations providing the separated solutions of the partial differential equation satisfied by the function $\phi(x, y) \equiv X(x)\, Y(y)$

$$\frac{\partial^2 \phi}{\partial x^2} + \frac{\partial^2 \phi}{\partial y^2} + 6\frac{\partial \phi}{\partial x} + 9\phi = 0.$$

Answers

1. $x^2 D^2 y + x Dy + (x^2 - \tfrac{1}{4})y = 0.$
2. $Du + 2u + 3v = 0, \quad Dv + 3u + 2v = 0.$
3. $d^2 u/d\theta^2 + u = f/h^2 u^2.$
4. $D_x^2 X + 6 D_x X + (9 - n^2)X = 0, \quad D_y^2 Y + n^2 = 0.$

Chapter 2

First Order Differential Equations

2.1 Separable Equations If, in the first order differential equation $Dy = f(x, y)$, the function $f(x, y)$ can be factorized into two factors, one a function of x alone and the other a function of y alone, namely $f(x, y) \equiv g(x) h(y)$, then the equation is a *separable equation*.

Equations, not immediately separable, may be transformed to have this property as follows:

When $f(x, y) \equiv g(ax + by + c)$, let $z = ax + by + c$.

When $f(x, y) \equiv g(y/x)$, let $y = vx$.

When $f(x, y) \equiv g[(ax + by + c)/(Ax + By + C)]$, change the origin to the point of intersection of the two lines $ax + by + c = 0$, $Ax + By + C = 0$, $aB \neq Ab$.

Problem 2.1 Solve the equation
$$Dy = x^2 + xe^{-y} - x^2 e^{-y} - x.$$

Solution. This is separable since inspection shows that the right-hand side may be factorized:
$$Dy = (x^2 - x)(1 - e^{-y}).$$

Prepared for integration, this gives
$$dy/(1 - e^{-y}) = (x^2 - x)\, dx,$$
$$e^y\, dy/(e^y - 1) = (x^2 - x)\, dx.$$

Upon integration
$$\log(e^y - 1) = \tfrac{1}{3}x^3 - \tfrac{1}{2}x^2 + A, \qquad y > 0.$$

Expressed explicitly,
$$y = \log\left[1 + \exp(\tfrac{1}{3}x^3 - \tfrac{1}{2}x^2 + A)\right]. \qquad \square$$

Problem 2.2 Solve the equation
$$Dy = \cos(x + y - 2),$$

so that $y = \tfrac{1}{2}\pi$ when $x = 2$.

Solution. Introduce the standard substitution $z = x + y - 2$, yielding
$$\frac{dz}{dx} = 1 + \frac{dy}{dx} = 1 + \cos z.$$

Prepared for integration, this equation separates immediately to

7

$$dx = \frac{dz}{1+\cos z} = \frac{dz}{2\cos^2 \frac{1}{2}z} = \tfrac{1}{2}\sec^2 \tfrac{1}{2}z\, dz.$$

Integration yields

$$x + A = \tan \tfrac{1}{2}z$$
$$= \tan \tfrac{1}{2}(x+y-2).$$

Now $y = \tfrac{1}{2}\pi$ when $x = 2$, so $A = -1$ (since $\tan \tfrac{1}{4}\pi = 1$), giving

$$y = 2\tan^{-1}(x-1) - x + 2. \qquad \square$$

Problem 2.3 Solve the equation

$$xy\, Dy = 2x^2 + 3xy + 2y^2,$$

so that $y = 1$ when $x = 1$.

Solution. Since $f(x, y) \equiv (2x^2 + 3xy + 2y^2)/xy \equiv 2(x/y) + 3 + 2(y/x)$, we introduce the standard substitution $y = vx$. Then

$$dy/dx = v + dv/dx = 2v^{-1} + 3 + 2v,$$

which separates to

$$\frac{dx}{x} = \frac{dv}{2v^{-1} + 3 + v} = \frac{v\, dv}{v^2 + 3v + 2} = \left(\frac{2}{v+2} - \frac{1}{v+1}\right) dv$$

when arranged in terms of partial fractions. Integration yields

$$\log x = 2\log(v+2) - \log(v+1) + \text{constant}.$$

Hence

$$x = \frac{A(v+2)^2}{v+1} = \frac{A(y+2x)^2}{x(y+x)};$$

(note that when logarithms are removed, a multiplicative constant replaces an additive constant). Since $y = 1$ when $x = 1$.

$$9x^2(y+x) = 2(y+2x)^2.$$

There is little point in solving this explicitly for y in terms of x as a quadratic equation in y.

Problem 2.4 Find the general solution of the equation

$$\frac{dy}{dx} = 36\left[\frac{x+y+1}{11x+y-19}\right]^2.$$

Solution. The two lines $x+y+1 = 0$, $11x+y-19 = 0$ intersect at $x = 2$, $y = -3$. Hence introduce the standard substitution $X = x-2$, $Y = y+3$:

$$\frac{dY}{dX} = 36\left[\frac{X+Y}{11X+Y}\right]^2.$$

Introduce the second standard substitution $Y = vX$:

$$\frac{dY}{dX} = v + X\frac{dv}{dX} = 36\left[\frac{1+v}{11+v}\right]^2,$$

$$\frac{dX}{X} = \frac{dv}{36[(1+v)/(11+v)]^2 - v} = \frac{-(11+v)^2\,dv}{v^3 - 14v^2 + 49v - 36}$$

$$= \left[-\frac{6}{v-1} + \frac{15}{v-4} - \frac{10}{v-9}\right]dv$$

when expressed in terms of partial fractions. Integrating,

$$\log X = -6\log(v-1) + 15\log(v-4) - 10\log(v-9) + \text{constant},$$
$$X = A(v-4)^{15}/(v-1)^6(v-9)^{10},$$

i.e. $\qquad (Y-X)^6(Y-9X)^{10} = A(Y-4X)^{15}.$

In terms of x and y,

$$(y-x+5)^6(y-19x+21)^{10} = A(y-4x+11)^{15}. \qquad \square$$

2.2 Exact Equations A first order differential equation $Dy = f(x, y)$ may often be expressed in the alternative form

$$m(x,\ y)\,dx + n(x, y)\,dy = 0. \qquad (1)$$

If there exists a function $F(x, y)$ such that the first order increment in its Taylor expansion,

$$dF(x, y) \equiv \frac{\partial F}{\partial x}\,dx + \frac{\partial F}{\partial y}\,dy,$$

is identical with $m(x, y)dx + n(x, y)dy$, then equation (1) is *exact*. The equation is exact when

$$\partial m/\partial y \equiv \partial n/\partial x;$$

its solution is the family of curves given by $F(x, y) = \text{constant}$. If (1) is not exact, it may be rendered exact by multiplication by a suitable *integrating factor* $J(x, y)$. In elementary work, a hint is necessary to find $J(x, y)$. $\quad\square$

Problem 2.5 Solve the equation

$$(2x+y)\,dx + (x+2y)\,dy = 0.$$

Solution. The equation is obviously exact, since it can be written in the form

$$\frac{\partial(x^2+xy)}{\partial x}\,dx + \frac{\partial(xy+y^2)}{\partial y}\,dy = 0,$$

or
$$\frac{\partial(x^2+xy+y^2)}{\partial x}dx+\frac{\partial(x^2+xy+y^2)}{\partial y}dy = 0,$$

where the inserted terms y^2 and x^2 respectively yield zero when differentiated. Hence
$$d(x^2+xy+y^2) = 0$$
$$x^2+xy+y^2 = A. \qquad \square$$

Problem 2.6 Solve the equation
$$(3x^2y+y^3)\,dx+(x^3+3y^2x+1)\,dy = 0.$$

Solution. Writing $m = 3x^2y+y^3$, $n = x^3+3y^2x+1$, we see that
$$\partial m/\partial y = 3x^2+3y^2 = \partial n/\partial x,$$
so the equation is exact. There exists a function $F(x, y)$ such that
$$\partial F/\partial x = 3x^2y+y^3, \qquad \partial F/\partial y = x^3+3y^2x+1.$$

Partial integration of this first relation with respect to x yields
$$F = x^3y+y^3x+Y(y),$$
where $Y(y)$ is an unknown function of y. Then redifferentiation yields
$$\partial F/\partial y = x^3+3y^2x+dY/dy = x^3+3y^2x+1.$$
Hence $dY/dy = 1$ and $Y = y$. The solution is finally
$$F \equiv x^3y+y^3x+y = A. \qquad \square$$

Problem. 2.7 If $(Mdx+Ndy)e^{g(x)} = 0$ is an exact equation, verify that
$$\frac{\partial M}{\partial y} = \frac{\partial N}{\partial x}+N\frac{dg}{dx}.$$

Prove that an integrating factor can be deduced for the equation
$$M\,dx+N\,dy = 0$$
if $(\partial M/\partial y-\partial N/\partial x)/N$ is a function of x alone.
Solve the equation
$$(3xy-2ay^2)\,dx+(x^2-2axy)\,dy = 0.$$

Solution. The condition for the given equation to be exact is
$$\frac{\partial}{\partial y}(Me^{g(x)}) = \frac{\partial}{\partial x}(Ne^{g(x)}),$$
i.e.
$$e^g\frac{\partial M}{\partial y} = e^g\frac{\partial N}{\partial x}+Ne^g\frac{dg}{dx},$$

reducing to the given condition.

10

Conversely, if $(\partial M/\partial y - \partial N/\partial x)/N$ is a function of x alone, we may evaluate

$$g(x) = \int \frac{1}{N}\left(\frac{\partial M}{\partial y} - \frac{\partial N}{\partial x}\right) dx,$$

yielding the factor e^g that renders the equation exact.

In the given case, $M = 3xy - 2ay^2$, $N = x^2 - 2axy$, with

$$\frac{1}{N}\left(\frac{\partial M}{\partial y} - \frac{\partial N}{\partial x}\right) = \frac{3x - 4ay - 2x + 2ay}{x^2 - 2axy} = \frac{1}{x};$$

hence

$$g(x) = \int dx/x = \log x,$$

yielding the integrating factor $e^g \equiv e^{\log x} = x$. Thus the equation

$$(3x^2y - 2axy^2)\,dx + (x^3 - 2ax^2y)\,dy = 0$$

is exact, with the obvious solution

$$x^3y - ax^2y^2 = A. \qquad \square$$

2.3 Linear Equations   The equation

$$Dy + f(x)\,y = g(x)$$

represents the standard form of the *first order linear equation*. The *integrating factor* $J \equiv \exp\left[\int f(x)\,dx\right]$ (the integral involved needs no constant of integration) reduces the equation to

$$D(Jy) = Jg,$$

which may be integrated to yield y. Integration by parts is often necessary, and preparatory care must be exercised in the process as one of the following examples shows. Sometimes an equation may be rendered linear when dx/dy is used rather than dy/dx, y now being the independent variable. Sometimes a change of variable, such as in Bernoulli's equation, may reduce a given equation to linear form.

Problem 2.8   Find the general solution of the linear equation

$$\cos x\, Dy + (\sin x + \cos x)\,y = \sin 2x + 2.$$

Solution. After division by $\cos x$ (to render the coefficient of Dy equal to unity), the integrating factor J is

$$J = \exp\left[\int (\tan x + 1)\,dx\right] = \exp(-\log \cos x + x) = e^x \sec x.$$

The equation becomes, after multiplication by J:

$$D(ye^x \sec x) = e^x \sec x \frac{\sin 2x + 2}{\cos x} = 2e^x(\tan x + \sec^2 x).$$

Integration yields

$$ye^x \sec x = 2 \int e^x(\tan x + \sec^2 x)\,dx + A$$
$$= 2e^x \tan x + A,$$

by inspection. Hence the general solution is

$$y = 2\sin x + Ae^{-x}\cos x. \qquad \square$$

Problem 2.9   Find the general solution of the equation

$$y^2(y^6 - x^2)\,dy/dx = 2x.$$

Solution. The equation is not linear in its present form, but a certain judicious rearrangement yields a more familiar form. If x is the dependent variable,

$$2x\,dx/dy = y^2(y^6 - x^2).$$

If we write $u = x^2$, then $du/dy = 2x\,dx/dy$, yielding

$$du/dy + y^2 u = y^8.$$

This is now a linear equation. The integrating factor is

$$J = \exp\left(\int y^2\,dy\right) = \exp\left(\tfrac{1}{3}y^3\right),$$

$$d[u\exp\left(\tfrac{1}{3}y^3\right)]/dy = y^8 \exp\left(\tfrac{1}{3}y^3\right).$$

Integration by parts requires the integral of the product $y^2 \exp\left(\tfrac{1}{3}y^3\right)$. Hence

$$u\exp\left(\tfrac{1}{3}y^3\right) = \int y^6 \cdot y^2 \exp\left(\tfrac{1}{3}y^3\right)\,dy + A$$
$$= y^6 \exp\left(\tfrac{1}{3}y^3\right) - 6\int y^5 \exp\left(\tfrac{1}{3}y^3\right)\,dy + A$$
$$= y^6 \exp\left(\tfrac{1}{3}y^3\right) - 6\int y^3 \cdot y^2 \exp\left(\tfrac{1}{3}y^3\right)\,dy + A$$
$$= y^6 \exp\left(\tfrac{1}{3}y^3\right) - 6y^3 \exp\left(\tfrac{1}{3}y^3\right) + 18\int y^2 \exp\left(\tfrac{1}{3}y^3\right)\,dy + A$$
$$= (y^6 - 6y^3 + 18)\exp\left(\tfrac{1}{3}y^3\right) + A.$$

Hence $\qquad u = x^2 = y^6 - 6y^3 + 18 + A\exp\left(-\tfrac{1}{3}y^3\right).$

This equation cannot be solved explicitly for y in terms of x. $\qquad \square$

Problem 2.10   Solve the equation

$$2x\,Dy + y = 2x^2(x+1)\,y^3.$$

Solution. The left-hand side is characteristic of a first order linear equation, but the right-hand side equals a function of x multiplied by a power of y. Such equations are known as *Bernoulli's equations*.

Divide by y^3 and place $u = y^{-2}$. (If y^n appears on the right-hand side,

12

we divide by y^n and place $u = y^{-(n-1)}$.) Then $Du = -2y^{-3} Dy$, yielding
$$-x\,Du+u = 2x^2(x+1),$$
$$Du-u/x = -2x(x+1).$$
The integrating factor is
$$J = \exp\left(-\int dx/x\right) = 1/x,$$
so
$$D(u/x) = -2(x+1).$$
Integration yields
$$u/x = A-(x+1)^2,$$
or
$$y^{-2} = u = Ax-x(x+1)^2. \qquad \square$$

2.4 Equations of Higher Degree

The equation may either be solved for Dy and then integrated, or special methods may be used as in Clairaut's equation and the Riccati equation. If the family of solutions is expressed in the form $f(x, y, A) = 0$, where A denotes the arbitrary constant of integration yielding distinct members of the family, and if A can be eliminated between the equations $f(x, y, A) = 0$, $\partial f(x, y, A)/\partial A = 0$, then the resulting curve is known as the envelope of the family. All member curves of the family touch the envelope, which is known as a *singular solution* of the equation when it satisfies the equation.

Problem 2.11 Solve the equation of the third degree
$$(Dy)^3 - 6x(Dy)^2 + 11x^2\,Dy - 6x^3 = 0.$$
Solution. This is a cubic equation in Dy, and may be factorized thus:
$$(Dy-x)(Dy-2x)(Dy-3x) = 0.$$
There are thus three independent equations $Dy = x$, $Dy = 2x$, $Dy = 3x$, integrating to
$$y = \tfrac{1}{2}x^2+A, \quad y = x^2+B, \quad y = \tfrac{3}{2}x^2+C,$$
representing three families of parabolae, all with their axes along the positive y-axis.

Problem 2.12 Solve the Clairaut equation
$$y = x\,Dy-(27/256)(Dy)^4.$$
Solution. In standard notation, we write $Dy \equiv p$:
$$y = xp-(27/256)p^4.$$
Differentiation with respect to x yields
$$p = p+x\,Dp-(27/64)p^3\,Dp,$$
i.e.
$$Dp = 0 \quad \text{or} \quad x-(27/64)p^3 = 0.$$

13

B

In the former case, $p = A$, so
$$y = Ax - (27/256) A^4,$$
representing a family of straight lines whose intercept on the y-axis depends on the gradient of the line.

In the latter case, $p = 4x^{\frac{1}{3}}/3$, leading to
$$y = x(\tfrac{4}{3}x^{\frac{1}{3}}) - (27/256)(\tfrac{4}{3}x^{\frac{1}{3}})^4 = x^{\frac{4}{3}}.$$

This is the singular solution, and is, in fact, the envelope of the first general solution. $\qquad\square$

Problem 2.13 Solve the Riccati equation
$$y^2 + Dy - (12/x^2) = 0.$$

(The first two terms characterize a Riccati equation, the last term being a function of x).

Solution. Let $y = (Du)/u$; then
$$Dy = \frac{D^2u}{u} - \frac{(Du)^2}{u^2},$$

and the given equation becomes
$$\left(\frac{Du}{u}\right)^2 + \left(\frac{D^2u}{u} - \frac{(Du)^2}{u^2}\right) - \frac{12}{x^2} = 0,$$
or
$$D^2u - 12u/x^2 = 0.$$

Taking $u = x^n$, we see that $n(n-1) - 12 = 0$, i.e. $n = 4, -3$. Hence $u = Ax^4 + Bx^{-3}$, and
$$y = \frac{4Ax^3 - 3Bx^{-4}}{Ax^4 + Bx^{-3}} = \frac{4Cx^7 - 3}{Cx^8 + x},$$

where only one arbitrary constant $C\,(= A/B)$ is involved. $\qquad\square$

Preliminary work may often be carried out by the *method of isoclines*. The equation is expressed in the form
$$Dy = f(x, y).$$

The equation $f(x, y) = m$, where m is a given constant, defines a curve C at all points of which the integral curve of the differential equation passing through such a point has gradient m. The curve C is sketched on the (x, y)-plane, and at a large number of points on it small line segments are drawn with gradient m. This is done for many values of m, after which the integral curves may be sketched through the line segments.

14

Problem 2.14   Sketch the integral curves of the equation
$$Dy = y^2 - x^2.$$

Solution. In this case, an analytical solution is possible. If we place $z = -y$, a Riccati equation is obtained; let $(Du)/u = z$, leading to $D^2u = x^2u$, whose solution is known to be (see page 91)
$$u = x^{\frac{1}{2}}[AJ_{\frac{1}{4}}(\tfrac{1}{2}ix^2) + BJ_{-\frac{1}{4}}(\tfrac{1}{2}ix^2)]$$
in terms of Bessel functions of order $\pm\frac{1}{4}$. Hence $y = -(Du)/u$ is determined.

It is not possible to visualize this curve without the extensive use of tables of Bessel functions, but the method of isoclines yields a rough outline immediately.

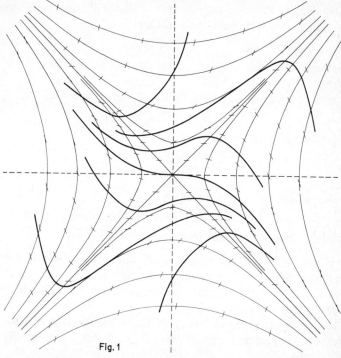

Fig. 1

The hyperbolae $m = y^2 - x^2$ are plotted for $m = \pm 4, \pm\frac{9}{4}, \pm 1, \pm\frac{1}{4}, 0$ (cutting the axes at points with coordinates $\pm 2, \pm\frac{3}{2}, \pm 1, \pm\frac{1}{2}, 0$ respectively) as in the diagram. Line segments on these hyperbolae are then sketched in with gradients $\pm 4, \pm\frac{9}{4}, \pm 1, \pm\frac{1}{4}, 0$ respectively. The sketch of the integral curves can then be made as shown, by a method similar to the plotting of magnetic lines of force in a plane from specific directions at a large number of discrete points.   □

15

2.5 Applications

Problem 2.15   At time $t = 0$, a battery of constant e.m.f. E is switched into a circuit containing a resistance R and a capacitance C. Find the current i and the charge q on the condenser at time t.

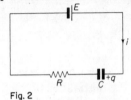

Fig. 2

Solution. Let i denote the current flowing into the plate containing the charge q at time t; then $i = Dq$, where $D \equiv d/dt$. The equation governing the passage of current through the resistance is

$$E - q/C = iR,$$

the left-hand side being the e.m.f. in the circuit in the direction of the current.

This may now be expressed as a first order linear equation in q:

$$R\,Dq + q/C = E,$$

possessing the solution

$$q = EC + Ae^{-t/CR}$$
$$= EC(1 - e^{-t/CR})$$

since $q = 0$ when $t = 0$. The current is then

$$i = Dq = (E/R)\,e^{-t/CR},$$

representing exponential decay from the initial value E/R.   □

Problem 2.16   Find and integrate the differential equation representing the families of curves such that the rectangular axes Ox and Oy cut off equal intercepts from the tangent and normal at any point P of a curve.

Solution. Draw the curve with positive gradient in the first quadrant, the concave side being upwards (that is, the second derivative being also positive). The tangent at P cuts Ox and Oy in S and T respectively; the

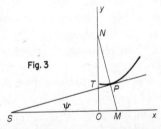

Fig. 3

16

normal at P cuts Ox and Oy at M and N respectively. It is given that $ST = MN$. If $\angle PSx = \psi$, we have

$$SP - TP = NP + PM,$$

the intercept ST lying in the second quadrant. Hence

$$\frac{y}{\sin\psi} - \frac{x}{\cos\psi} = \frac{x}{\sin\psi} + \frac{y}{\cos\psi},$$

or

$$\tan\psi = \frac{dy}{dx} = \frac{y - x}{y + x}.$$

To solve this equation, let $y = vx$. Then

$$Dy = x\,Dv + v = \frac{v - 1}{v + 1},$$

whence

$$\frac{(1 + v)\,dv}{1 + v^2} = -\frac{dx}{x}.$$

This integrates to

$$\tan^{-1} v + \tfrac{1}{2}\log(1 + v^2) = -\log x + \text{constant},$$

or

$$\tan^{-1}(y/x) + \tfrac{1}{2}\log(x^2 + y^2) = A.$$

If the intercept ST lies in the fourth quadrant, a slightly different equation is obtained with a similar solution but with x and y interchanged in position. ∎

Problem 2.17 Electric point charges e, $-2e$, e are placed at the points $z = -a$, 0, a respectively along the z-axis. In the limit as $a \to 0$ and $e \to \infty$ such that $ea^2 \equiv q$ remains a constant, it is given that the electric potential ϕ attains the limiting form

$$4\pi\varepsilon_0\,\phi = q(3\cos^2\theta - 1)/r^3.$$

Find the equation of the lines of force due to this limiting source at the origin.

Solution. The field components in polar coordinates are given by

$$E_r = -\partial\phi/\partial r, \qquad E_\theta = -\partial\phi/r\,\partial\theta.$$

The direction of a line of force at a point P is given firstly by the geometrical increments δr, $r\,\delta\theta$, and secondly by the field components E_r, E_θ. Equating their ratios, we obtain

$$\frac{dr}{r\,d\theta} = \frac{\partial\phi/\partial r}{\partial\phi/r\,\partial\theta} = \frac{[3(3\cos^2\theta - 1)]/r^4}{(6\cos\theta\sin\theta)/r^4} = \frac{3\cos^2\theta - 1}{2\cos\theta\sin\theta}.$$

17

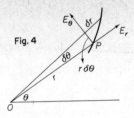

Fig. 4

This separable equation of the first order may be arranged for integration as follows:

$$\frac{2\,dr}{r} = \frac{3\cos^2\theta - 1}{\cos\theta\,\sin\theta}\,d\theta$$

$$= \frac{2\cos^2\theta - \sin^2\theta}{\cos\theta\,\sin\theta}\,d\theta = (2\cot\theta - \tan\theta)\,d\theta.$$

Integrating,

$$2\log r = 2\log\sin\theta + \log\cos\theta + \text{constant},$$

or

$$r^2 = A\sin^2\theta\cos\theta.$$
□

Problem 2.18 The differential equation representing a family of plane curves is

$$Dy = (y^2 - x^2)/2\,xy.$$

Integrate this equation and find the equation of the *orthogonal trajectories*.

Solution. The given equation is either homogeneous or it is a linear equation in $u \equiv y^2$, for

$$2y\,Dy - x^{-1}y^2 = -x,$$

i.e.

$$Du - x^{-1}u = -x.$$

The integrating factor is $\exp(\int x^{-1}\,dx) = 1/x$, so

$$D(u/x) = -1,$$

integrating to

$$u/x = -x + 2A,$$

i.e.

$$u \equiv y^2 = -x^2 + 2Ax,$$

$$x^2 + y^2 - 2Ax = 0.$$

This is a family of circles with centre at the point $(A, 0)$ and radius $|A|$.

At any point $P\,(x, y)$ on a curve with gradient $\tan\psi$, the new gradient of the orthogonal trajectory must be $-1/\tan\psi$, since the two curves intersect orthogonally at P. Hence

$$-\frac{1}{Dy} = \frac{y^2 - x^2}{2xy}$$

is the differential equation representing the family of orthogonal trajectories. This is

$$\frac{dx}{dy} = \frac{x^2 - y^2}{2xy},$$

identical with the original equation, but with x and y interchanged. Its
solution is therefore

$$x^2 + y^2 - 2By = 0,$$

a family of circles with centre at $(0, B)$ and radius $|B|$. $\square$

EXERCISES

1.   Solve $y\,Dy = \sin x\,e^{x+2y}$.

2.   Solve $xy\,Dy = 2x^2 + 3xy + 2y^2$, such that the solution passes through
the point $(1, 0)$.

3.   Solve $(4x^2 + 2xy + 6y)\,dx + (2x^2 + 9y + 3x)\,dy = 0$, given that an integrating factor of the form $(x + y)^n$ exists.

4.   Solve the equation $x\,Dy + (x - 1)\,y = x^3$, such that the solution passes
through the point $(1, 1)$.

5.   Solve the equation $y = x\,Dy - \exp(Dy)$, and find the singular solution.

6.   The normal at a point $P(x, y)$ on a plane curve cuts the x- and y-axes
at Q and R respectively on opposite sides of P; it is given that
$RP/RQ = x^2/y^2$. Find the curve satisfying these conditions passing
through the point $(1, 1)$.

7.   Two point charges e, $-e$ are separated by the distance $2a$; as $a \to 0$
and $e \to \infty$ such that the product ae remains constant, show that the
potential in the limit is proportional to $r^{-2}\cos\theta$. Show that the equation
of the lines of force is $r = A\sin^2\theta$ in polar coordinates.

Answers

1. $-\frac{1}{4}(1 + 2y)e^{-2y} = \frac{1}{2}(\sin x - \cos x)e^x + A$.

2. $(y + 2x)^2 = 4^2(y + x)$.

3. $n = 1$; $x^4 + 2x^3y + 6y^2x + 3x^2y + x^2y^2 + 3y^3 = A$.

4. $y = x^2 - x + xe^{-x}$.

5. $y = Ax - e^A$; $y = x\log x - x$.

6. $y^2/x^2 = 1 - 2\log x$.

Chapter 3

Second Order Differential Equations

Many of the methods illustrated in this chapter relating to second order differential equations may be applied to equations of higher order n; clarity is maintained by restricting the order to $n = 2$ whenever possible.

The general *inhomogeneous* linear equation of order 2,

$$f(x)\,D^2 y + g(x)\,Dy + h(x)\,y = j(x), \tag{1}$$

does not possess any explicit form of solution in terms of quadratures, namely, the equation cannot be rearranged for general functions f, g, h, j into a form where integration is possible. The existence of solutions (although not expressible in terms of standardized functions) is guaranteed by *existence theorems* when f, g, h, j satisfy suitable requirements. Analytic solutions are possible for a wide range of functions f, g, h, j. For other equations, approximate analytical solutions may be found, while in other cases, numerical methods including the use of a computer are the only devices available.

The general solution of (1) consists of the sum of the *particular integral* (any one solution of the inhomogeneous equation) and the *complementary function* (the general solution containing two arbitrary constants of the equation with the inhomogeneous term $j(x)$ replaced by zero). Some procedures evaluate these two parts separately, while other methods yield the two parts of the general solution simultaneously without recourse to separate calculations. The Laplace transform method introduces the initial conditions *ab initio*, and does not proceed via the general solution.

3.1 Homogeneous Equations with Constant Coefficients If α and β are the roots of the *auxiliary equation*

$$f\lambda^2 + g\lambda + h = 0,$$

where f, g, h are *constants*, then the general solution of $fD^2 y + gDy + hy = 0$ is

$$y = Ae^{\alpha x} + Be^{\beta x} \qquad (\alpha \neq \beta),$$
$$y = (A + Bx)\,e^{\alpha x} \qquad (\alpha = \beta).$$

If $h = 0$ one root vanishes, and $y = A$ is the contribution to the general solution from this root.

In an nth order equation, if α is an r-fold repeated root of the corresponding auxiliary polynomial equation, then the general solution will contain the group of terms

$$(A_1+A_2x+A_3x^2+\ldots A_r x^{r-1})\,e^{\alpha x}.$$

Problem 3.1 Solve the homogeneous equation
$$D^2y-12Dy+35y = 0.$$

Solution. The auxiliary equation is
$$\lambda^2-12\lambda+35 = 0$$
with roots 5, 7. Hence
$$y = Ae^{5x}+Be^{7x}. \qquad\square$$

Problem 3.2 Solve the third order equation
$$D^3y-27Dy+54 = 0.$$

Solution. The auxiliary cubic equation $\lambda^3-27\lambda+54 = 0$ factorizes to
$$(\lambda-3)^2\,(\lambda+6) = 0,$$
which possesses roots 3, 3, -6. Hence
$$y = (A+Bx)\,e^{3x}+Ce^{-6x}. \qquad\square$$

Problem 3.3 Solve the equation
$$D^2y-4Dy+13y = 0$$
in real terms, where complex roots of the auxiliary quadratic equation are involved.

Solution. The auxiliary quadratic $\lambda^2-4\lambda+13 = 0$ has roots
$$\lambda = \tfrac{1}{2}[4\pm\sqrt{(16-52)}] = 2\pm3i.$$
Hence
$$
\begin{aligned}
y &= Ae^{(2+3i)x}+Be^{(2-3i)x}\\
&= Ae^{2x}(\cos 3x+i\sin 3x)+Be^{2x}(\cos 3x-i\sin 3x)\\
&= e^{2x}[(A+B)\cos 3x+(A-B)\,i\sin 3x]\\
&= e^{2x}(C\cos 3x+D\sin 3x),
\end{aligned}
$$
where C and D remains arbitrary constants. C and D are real for real solutions, implying that A and B must be complex when the complex exponential forms are used. $\qquad\square$

Without use of complex numbers, the equation
$$f\,D^2y+g\,Dy+hy = 0$$
(with f, g, h constants) may be solved by introducing the elementary change of dependent variable $y = e^{-(g/2f)x}v$, where v is the new dependent variable.

Problem 3.4   Solve the equation
$$D^2y - 6Dy + 13y = 0,$$
where the auxiliary quadratic has complex roots.

Solution. Introduce the change of variable $y = e^{(6/2)x}v \equiv e^{3x}v$. Then
$$Dy = e^{3x}Dv + 3e^{3x}v,$$
$$D^2y = e^{3x}D^2v + 6e^{3x}Dv + 9e^{3x}v.$$

Upon substitution into the given equation,
$$(D^2v + 6Dv + 9v) - 6(Dv + 3v) + 13v = 0,$$
$$D^2v + 4v = 0,$$

the equation of simple harmonic motion. We have
$$v = A\cos 2x + B\sin 2x,$$
$$y = e^{3x}(A\cos 2x + B\sin 2x). \qquad \square$$

3.2 Particular Integrals   Particular integrals may always be found when f, g, h are constants, and when j is a polynomial in x, an exponential $e^{\alpha x}$ (or a sum of such functions such as sinh αx, cosh αx), a trigonometric function of the form sin αx or cos αx (or functions that may be reduced to sums of such forms, such as $\sin^3 \alpha x$), or products of these functions.

Often the method is advocated by which the particular integral is ascertained by intelligent guessing, with unknown constant coefficients to be found by direct substitution. The D operator method, however, is a rigorously based method, and produces the particular integrals without guesswork. It is based on three simple rules of manipulation.

(i) When $j(x)$ is a polynomial of order n in x, expand the *operator* $1/(f D^2 + g D + h)$ in the formal solution

$$y = \frac{1}{f D^2 + g D + h}j(x)$$

by the binomial theorem to the nth power of D. D^r in the numerator then means differentiate $j(x)$ r times, while D^r in the denominator means integrate $j(x)$ r times.

(ii) When $j(x)$ is an exponential function $e^{\alpha x}$, replace D by α unless $f\alpha^2 + g\alpha + h = 0$.

(iii) When $f\alpha^2 + g\alpha + h = 0$, of if $j(x) = e^{\alpha x}$ multiplied by a polynomial, replace D by $D + \alpha$, and remove the factor $e^{\alpha x}$ to the left of the new operator, allowing this to operate on terms remaining on its right.

For this purpose, sines and cosines are expressed in terms of complex exponential functions. If all coefficients such as f, g, h are real, $\cos \alpha x$ and $\sin \alpha x$ may be replaced by $\operatorname{Re} e^{i\alpha x}$ and $\operatorname{Im} e^{i\alpha x}$ respectively.

Under certain circumstances, the D operator method may be used even when $j(x)$ is of a form different from those specified.

Problem 3.5 Find the general solution of the equation
$$D^3 y - D^2 y - Dy + y = 7 - 6x - 3x^2 + x^3.$$

Solution. The auxiliary equation is $\lambda^3 - \lambda^2 - \lambda + 1 = 0$ which factorizes to $(1+\lambda)(1-\lambda)^2 = 0$ with roots 1, 1, -1. Hence the complementary function is
$$y = (A + Bx)\, e^x + Ce^{-x}.$$

By rule (i), the particular integral is given by
$$y = \frac{1}{D^3 - D^2 - D + 1} (7 - 6x - 3x^2 + x^3)$$

$$= \frac{1}{(1+D)(1-D)^2} (7 - 6x - 3x^2 + x^3)$$

$$= \frac{1}{4}\left[\frac{1}{1+D} + \frac{1}{1-D} + \frac{2}{(1-D)^2} \right] (7 - 6x - 3x^2 + x^3) \quad \text{(in partial fractions)}$$

$$= \tfrac{1}{4}[(1 - D + D^2 - D^3 \ldots) + (1 + D + D^2 + D^3 \ldots)$$
$$\qquad + 2(1 + 2D + 3D^2 + 4D^3 \ldots)]\,(7 - 6x - 3x^2 + x^3)$$

$$= (1 + D + 2D^2 + 2D^3 + \ldots)\,(7 - 6x - 3x^2 + x^3)$$

$$= (7 - 6x - 3x^2 + x^3) + (-6 - 6x + 3x^2) + 2(-6 + 6x) + 2(6)$$

$$= 1 + x^3.$$

Hence the general solution is
$$y = 1 + x^3 + (A + Bx)\, e^x + Ce^{-x}. \qquad \square$$

Note. If the form of the particular integral is to be 'guessed', a polynomial of degree three must be chosen. Assume
$$y = a + bx + cx^2 + dx^3.$$
Substituting,
$$D^3 y - D^2 y - Dy + y = 6d - (2c + 6dx) - (b + 2cx + 3dx^2)$$
$$+ (a + bx + cx^2 + dx^3)$$

$$= (6d - 2c - b + a) + (-6d - 2c + b)\, x$$
$$+ (-3d + c)\, x^2 + dx^3$$

$$= 7 - 6x - 3x^2 + x^3.$$

Identifying coefficients, we have four equations with which to determine the four constants a, b, c, d:

$$6d - 2c - b + a = 7,$$
$$-6d - 2c + b = -6,$$
$$-3d + c = -3,$$
$$d = 1.$$

Working backwards from the last equation, we find that $d = 1$, $c = 0$, $b = 0$, $a = 1$, yielding $y = 1 + x^3$ as before.

Problem 3.6 Find the particular integral of the equation

$$(D^2 - 7D + 12)y = 14e^{-3x} + 4e^{5x}.$$

Solution. The fact that the numerical coefficients -3, 5 occurring in the index of the exponentials differ from the roots 3, 4 of the auxiliary quadratic, shows that rule (ii) may be used.

Using this rule for each exponential separately, we find

$$y = \frac{14}{D^2 - 7D + 12}e^{-3x} + \frac{4}{D^2 - 7D + 12}e^{5x}$$

$$= \frac{14}{(-3)^2 - 7(-3) + 12}e^{-3x} + \frac{4}{5^2 - 7.5 + 12}e^{5x}$$

$$= \tfrac{1}{3}e^{-3x} + 2e^{5x},$$

where D has been replaced by -3 and 5 respectively   □

Note. For each exponential function on the right-hand side, provided that it is distinct from any such term in the complementary function, we may assume a particular integral proportional to the exponential function. Thus for the factor $14e^{-3x}$, assume a particular integral ae^{-3x}. Substitution yields

$$a(9 + 21 + 12)e^{-3x} = 14e^{-3x},$$

or $a = \tfrac{1}{3}$, as before.

Problem 3.7 Find the particular integral of the equation

$$D^2 y - 7Dy + 12y = 4\cosh^2 2x.$$

Solution. Firstly, we express the right-hand side in terms of exponential functions:

$$4\cosh^2 2x = 2\cosh 4x + 2 = e^{4x} + e^{-4x} + 2.$$

We note that the auxiliary quadratic possesses roots 3, 4, one of which equals the coefficient of x in the index of the first exponential. Rule (iii)

24

must be used in this case, but rule (ii) must be used for e^{-4x}. We have

$$y = \frac{1}{D^2 - 7D + 12} e^{4x} + \frac{1}{D^2 - 7D + 12} e^{-4x} + \frac{1}{D^2 - 7D + 12} \cdot 2$$

$$= e^{4x} \frac{1}{(D+4)^2 - 7(D+4) + 12} \cdot 1 + \frac{1}{(-4)^2 - 7(-4) + 12} e^{-4x} + \frac{1}{12} \cdot 2$$

$$= e^{4x} \frac{1}{D^2 + D} \cdot 1 + \frac{1}{56} e^{-4x} + \frac{1}{6}$$

$$= e^{4x} \left(\frac{1}{D} - 1 \dots \right) 1 + \frac{1}{56} e^{-4x} + \frac{1}{6}$$

$$= e^{4x}(x - 1) + \frac{1}{56} e^{-4x} + \frac{1}{6},$$

where $1/D$ implies integration. $\qquad\square$

Note. When an exponential function on the right-hand side is identical with part of the complementary function, the particular integral assumed must be proportional to this exponential function multiplied by x.

Thus the term e^{4x} leads to the assumed particular integral axe^{4x}. Substitution yields

$$a(8e^{4x} + 16xe^{4x}) - 7a(e^{4x} + 4xe^{4x}) + 12axe^{4x} = e^{4x},$$

or $a = 1$. The contribution to the particular integral is thus xe^{4x}. In the previously given solution, the contribution was found to be $(x - 1)e^{4x}$; the additional term $-e^{4x}$ is irrelevant, since this can be combined with the similar term occurring in the complementary function.

Problem 3.8 Find the particular integral of the equation

$$D^2 y + 2Dy + 4y = 2 \sin^2 3x.$$

Solution. The right-hand side should be expressed in terms of complex exponential functions:

$$2 \sin^2 3x = 1 - \sin 6x = 1 - \text{Im } e^{6ix}.$$

Rules (i) and (ii) are used for these two parts:

$$y = \frac{1}{D^2 + 2D + 4} \cdot 1 - \text{Im} \frac{1}{D^2 + 2D + 4} e^{6ix}$$

$$= \tfrac{1}{4} - \text{Im} \frac{1}{(6i)^2 + 2(6i) + 4} e^{6ix}$$

$$= \tfrac{1}{4} - \text{Im} \frac{e^{6ix}}{12i - 32}$$

25

$$= \tfrac{1}{4} - \tfrac{1}{4} \operatorname{Im} \frac{(-3i-8)(\cos 6x + i \sin 6x)}{(-3i-8)(3i-8)}$$

$$= \tfrac{1}{4} + (1/292)(3\cos 6x + 8\sin 6x). \qquad \square$$

Note. Since the right-hand side of the equation, $1 - \sin 6x$, is distinct from any part of the complementary function, the particular integral corresponding to $-\sin 6x$ must be taken to be

$$a\sin 6x + b\cos 6x.$$

Substituting,

$$-36a\sin 6x - 36b\cos 6x + 12a\cos 6x - 12b\sin 6x$$

$$+ 4a\sin 6x + 4b\cos 6x = -\sin 6x.$$

Identifying coefficients of $\sin 6x$ and $\cos 6x$ respectively,

$$-36a - 12b + 4a = -1,$$

$$-36b + 12a + 4b = 0,$$

with solutions $a = 8/292$, $b = 3/292$ as before.

Problem 3.9   Find the particular integral of the equation

$$D^2 y - 5Dy + 4y = -4(x^2 + 1)e^{3x}.$$

Solution. Rule (iii) must be used to separate the exponential function from the remaining polynomial. We have

$$y = \frac{1}{D^2 - 5D + 4}(-4)(x^2 + 1)e^{3x}$$

$$= e^{3x}\frac{1}{(D+3)^2 - 5(D+3) + 4}(-4)(x^2 + 1)$$

$$= \tfrac{2}{3}e^{3x}\left[\frac{2}{1 - D} + \frac{1}{1 + \tfrac{1}{2}D}\right](x^2 + 1)$$

$$= \tfrac{2}{3}e^{3x}[2(1 + D + D^2 \dots) + (1 - \tfrac{1}{2}D + \tfrac{1}{4}D^2 \dots)](x^2 + 1)$$

$$= \tfrac{2}{3}e^{3x}(3 + \tfrac{3}{2}D + \tfrac{9}{4}D^2 + \dots)(x^2 + 1)$$

$$= e^{3x}(2x^2 + 2x + 5). \qquad \square$$

Note. The exponential factor e^{3x} on the right-hand side is distinct from any part of the complementary function. Hence the particular integral to be assumed must be e^{3x} multiplied by a polynomial of degree two:

$$y = (a + bx + cx^2)e^{3x}.$$

Substitution and division by e^{3x} yield

$$(9a + 9bx + 9cx^2 + 3b + 6cx + 3b + 6cx + 2c)$$

$$-5(3a + 3bx + 3cx^2 + b + 2cx) + 4(a + bx + cx^2) = -4(x^2 + 1).$$

26

Identifying coefficients,

$$9a + 3b + 3b - 15a - 5b + 4a + 2c = -4,$$
$$9b + 6c + 6c - 15b - 10c + 4b = 0,$$
$$9c - 15c + 4c = -4.$$

Hence, $c = 2$, $b = 2$, $a = 5$, as before. □

Problem 3.10 Find the general solution of the equation

$$D^2 y - 2Dy + 5y = e^x \cos 2x.$$

Solution. The complex roots of the auxiliary quadratic are $1 \pm 2i$, so the complementary function is

$$e^x(A \cos 2x + B \sin 2x).$$

One of these, $e^x \cos 2x$, is identical with the right-hand side of the equation; rule (iii) must therefore be used. The particular integral is given by

$$y = \mathrm{Re}\,\frac{1}{D^2 - 2D + 5}\,e^x e^{2ix}.$$

Rule (ii) cannot be used since the denominator vanishes when D is replaced by $1 + 2i$. We therefore replace D by $D + 1 + 2i$:

$$y = \mathrm{Re}\,e^x e^{2ix}\,\frac{1}{(D + 1 + 2i)^2 - 2(D + 1 + 2i) + 5} \cdot 1$$

$$= \mathrm{Re}\,e^x e^{2ix}\,\frac{1}{D^2 + 4iD} \cdot 1$$

$$= \mathrm{Re}\,e^x e^{2ix}\, x/4i = \tfrac{1}{4}xe^x \sin 2x.$$

The general solution is then given by

$$y = e^x[A \cos 2x + (B + \tfrac{1}{4}x) \sin 2x]. \qquad □$$

Note. The complementary function contains the term $e^x \cos 2x$, so the particular integral must be assumed to be

$$y = axe^x \cos 2x + bxe^x \sin 2x, \qquad a, b \text{ real}.$$

Substitution tends to be lengthy unless special devices connected with the operator D are employed. Thus

$$y = \mathrm{Re}\,axe^{(1+2i)x} + \mathrm{Im}\,bxe^{(1+2i)x};$$

$$(D^2 y - 2Dy + 5y) = \mathrm{Re}\,ae^{(1+2i)x}[(D + 1 + 2i)^2 - 2(D + 1 + 2i) + 5]\,x$$
$$+ \mathrm{Im}\,be^{(1+2i)x}[(D + 1 + 2i)^2 - 2(D + 1 + 2i) + 5]\,x$$

$$= \mathrm{Re}\,ae^{(1+2i)x}(4i) + \mathrm{Im}\,be^{(1+2i)x}(4i)$$

$$= -4ae^x \sin 2x + 4be^x \cos 2x$$

$$= e^x \cos 2x,$$

as given. Hence $a = 0$, $b = \tfrac{1}{4}$.

The D operator method fails for further functions $j(x)$ because the operator $1/(f\,D^2+g\,D+h)$ cannot be interpreted by suitable rules to yield the solution. In *special* cases, progress is possible.

Problem 3.11 Find the particular integral of the equation
$$D^3y-3D^2y+3Dy-y = e^x\log x.$$

Solution. We employ rule (iii), obtaining

$$y = \frac{1}{(D-1)^3}e^x\log x$$

$$= e^x(1/D^3)\log x$$

$$= e^x(1/D^2)\int\log x\,dx$$

$$= e^x(1/D^2)(x\log x-x) \qquad \text{(by parts)}$$

$$= e^x(1/D)(\tfrac{1}{2}x^2\log x-\tfrac{3}{4}x^2) \qquad \text{(by parts)}$$

$$= e^x(\tfrac{1}{6}x^3\log x-\tfrac{11}{36}x^3) \qquad \text{(by parts).} \qquad \square$$

3.3 Some Special Methods If the differential equation does not contain the independent variable x explicitly, a new equation can be formed with $v = dy/dx$ as the dependent variable and y as the independent variable. For example, Newton's law of motion will not contain the time t explicitly when the force is given as a function of position. The equation will take the form

$$f(y, Dy, D^2y,\ldots) = 0,$$

with $D \equiv d/dx$. If we write $v = Dy$, then

$$D^2y = \frac{dv}{dx} = \frac{dv}{dy}\frac{dy}{dx} = v\frac{dv}{dy},$$

$$D^3y = \frac{d}{dx}\left(v\frac{dv}{dy}\right) = \frac{d}{dy}\left(v\frac{dv}{dy}\right)\frac{dy}{dx} = v\frac{d}{dy}\left(v\frac{dv}{dy}\right).$$

Problem 3.12 In Cartesian coordinates, all circles in the plane are governed by the differential equation
$$D^3y[1+(Dy)^2]-3Dy(D^2y)^2 = 0.$$

Recover the equation of the general circle.

Solution. The first stage of the integration may be carried out directly, since the equation is, upon careful inspection, separable, D^2y being used on the left and Dy on the right. We have

28

$$\frac{D^3y}{D^2y} = \frac{3Dy\, D^2y}{1+(Dy)^2},$$

i.e.
$$\frac{d(D^2y)}{D^2y} = \frac{3Dy\, d(Dy)}{1+(Dy)^2}.$$

Integrating,
$$\log D^2y = \tfrac{3}{2}\log\left[1+(Dy)^2\right]+\text{constant},$$
$$D^2y = \left[1+(Dy)^2\right]^{\frac{3}{2}}\times\text{constant}$$
$$= r^{-1}\left[1+(Dy)^2\right]^{\frac{3}{2}}, \quad \text{say}.$$

Since x is absent explicitly, the next stage is facilitated by introducing $Dy = v$ and $D^2y = v\, dv/dy$:

$$v\, dv/dy = r^{-1}(1+v^2)^{\frac{3}{2}},$$
$$rv\, dv/(1+v^2)^{\frac{3}{2}} = dy.$$

Integrating,
$$-r/(1+v^2)^{\frac{1}{2}} = y-b.$$

Squaring, solving for $v = dy/dx$, and arranging for integration, we obtain
$$\frac{(y-b)\,dy}{\left[r^2-(y-b)^2\right]^{\frac{1}{2}}} = dx.$$

Integrating
$$-\left[r^2-(y-b)^2\right]^{\frac{1}{2}} = x-a,$$
$$(x-a)^2+(y-b)^2 = r^2. \qquad\qquad \square$$

Exceptionally, the left-hand side of the equation
$$f\, D^2y+g\, Dy+hy = j,$$

where f, g, h, j are functions of x, may be an exact differential coefficient. If
$$D^2f - Dg + h \equiv 0,$$

then the equation may be written in the form
$$D[f\, Dy+(g-Df)\, y] = j.$$

This possesses the first integral
$$f\, Dy+(g-Df)\, y = \int j\, dx + A,$$

a linear first order equation.

Problem 3.13 Solve the equation
$$\cos x\, D^2y+\sec x\, Dy+(\sec x\tan x+\cos x) = 2\sec^2 x\tan x.$$

Solution. We note that
$$D^2f - Dg + h = -\cos x - \sec x\tan x + (\sec x\tan x + \cos x) \equiv 0,$$

29

c

so the equation may be rearranged thus:
$$D[\cos x \, Dy + (\sec x + \sin x) \, y] = 2 \sec^2 x \tan x.$$
The first integral is
$$\cos x \, Dy + (\sec x + \sin x) \, y = \sec^2 x + A,$$
$$Dy + (\sec^2 x + \tan x) \, y = \sec^3 x + A \sec x.$$
The integrating factor is
$$\exp\left[\int (\sec^2 x + \tan x) \, dx\right] = \exp(\tan x - \log \cos x) = \sec x \exp(\tan x).$$
Hence
$$D[y \sec x \exp(\tan x)] = \sec^4 x \exp(\tan x) + A \sec^2 x \exp(\tan x),$$
which integrates to
$$y \sec x \exp(\tan x) = \int \sec^4 x \exp(\tan x) \, dx + A \exp(\tan x) + B,$$
$$y = 3 \cos x - 2 \sin x + \sin x \tan x + A \cos x + B \cos x \exp(-\tan x),$$
where the integral has been evaluated by the substitution $u = \tan x$ followed by integration by parts. $\qquad\square$

If $y = Y(x)$ is one integral of the equation
$$f \, D^2 y + g \, Dy + h y = 0,$$
then a second independent solution is given by
$$Y(x) \int^x \frac{\exp\left[-\int (g/f) \, dx\right]}{Y^2(x)} \, dx.$$

Problem 3.14 If $e^{\alpha x}$ is given to be a solution of the equation
$$x \, D^2 y - 2(1 + x) \, Dy + (2 + x) \, y = 0,$$
find α and the general solution of the equation.

Solution. Substitution yields
$$x\alpha^2 - 2(1 + x)\alpha + (2 + x) = 0,$$
$$x(1 - 2\alpha + \alpha^2) + (2 - 2\alpha) = 0$$
for all x. Clearly $\alpha = 1$, so e^x is one solution.

To find the second solution, either the above formula may be quoted, or we may use the method by which the formula is established. Let a second solution be given by
$$y = e^x v(x).$$
Then $Dy = e^x(D + 1) v$, $D^2 y = e^x(D + 1)^2 v$, yielding
$$x(D + 1)^2 v - 2(1 + x)(D + 1) v + (2 + x) v = 0,$$
$$x \, D^2 v - 2 Dv = 0.$$

The first integral is $Dv = Cx^2$ and the second integral is $v = Bx^3$. Hence the general solution is

$$y = Ae^x + Bx^3 e^x.$$

3.4 Variation of Parameters If $u(x)$ and $v(x)$ are two independent solution of the equation

$$D^2 y + g\, Dy + hy = 0,$$

then the solution of the inhomogeneous equation

$$D^2 y + g\, Dy + hy = j(x),$$

such that $y(a)$ and $Dy(a)$ both vanish, is

$$y = -u(x) \int_a^x \frac{v(t)\, j(t)\, dt}{W(t)} + v(x) \int_a^x \frac{u(t)\, j(t)\, dt}{W(t)},$$

where $W(t) = u(dv/dt) - (du/dt)v$. Arbitrary initial values at $x = a$ may be introduced by inserting the complementary function $A\, u(x) + B\, v(x)$.

Until confidence is gained in the use of this formula, it is suggested that the student's working should follow the actual method by which the result is derived.

Problem 3.15 Find the general solution of the equation

$$D^2 y + y = \sec x.$$

Solution. The D operator method is inapplicable in this case. The solutions of the equation $D^2 y + y = 0$ are $\sin x$ and $\cos x$. Generally, let

$$y = M \sin x + N \cos x,$$

where M and N are functions of x. Then

$$Dy = (DM)\sin x + (DN)\cos x + M \cos x - N \sin x.$$

We choose $(DM)\sin x + (DN)\cos x = 0$; then

$$D^2 y = (DM)\cos x - (DN)\sin x - M \sin x - N \cos x$$
$$= -M \sin x - N \cos x + \sec x$$

from the general equation. Hence

$$(DM)\cos x - (DN)\sin x = \sec x,$$

these two equations for DM and DN showing that $DM = 1$ and $DN = -\tan x$. It follows that $M = x$ and $N = \log \cos x$, both vanishing at $x = 0$. Thus the solution, such that y and Dy both vanish at $x = 0$, is given by

$$y = x \sin x + \cos x \log \cos x.$$

The addition of the complementary function yields the general solution
$$y = A \sin x + B \cos x + x \sin x + \cos x \log \cos x. \qquad \square$$

Problem 3.16   Solve the equation
$$x^3 D^3 y - 3x^2 D^2 y + 6x Dy - 6y = 60x^6.$$

Solution. The solution of the homogeneous equation may be found by trying the solution $y = x^n$ (since r is the power of x that multiplies $D^r y$). Then
$$n(n-1)(n-2) - 3n(n-1) + 6n - 6 = 0,$$
$$(n-1)(n-2)(n-3) = 0.$$

Hence three solutions are x, x^2, x^3.

The solution of the inhomogeneous equation by the method of variation of parameters is taken to be
$$y = xM + x^2 N + x^3 P,$$
where M, N, P are functions of x. Differentiation yields
$$Dy = x\,DM + x^2\,DN + x^3\,DP + M + 2xN + 3x^2 P.$$
Choosing
$$x\,DM + x^2\,DN + x^3\,DP = 0, \qquad (1)$$
$$D^2 y = DM + 2x\,DN + 3x^2\,DP + 2N + 6xP.$$
Choosing
$$DM + 2x\,DN + 3x^2\,DP = 0, \qquad (2)$$
$$D^3 y = 2\,DN + 6x\,DP + 6P.$$

Substitution into the differential equation yields
$$2x^3\,DN + 6x^4\,DP = 60x^6. \qquad (3)$$

The solution of equations (1), (2), (3) is
$$DM = 30x^4, \quad DN = -60x^3, \quad DP = 30x^2,$$
and integration gives
$$M = 6x^5, \quad N = -15x^4, \quad P = 10x^3,$$
all vanishing at $x = 0$. Hence the solution is
$$y = (6 - 15 + 10)x^6 = x^6,$$
such that y, Dy, $D^2 y$ all vanish at $x = 0$. $\qquad \square$

3.5 Change of Variable   The general homogeneous equation
$$f(x) D^2 y + g(x) Dy + h(x) y = 0$$
does not possess solutions that may be expressed explicitly in terms of quadratures of functions involving f, g, h. Formally, if f, g and h are expressed as power series in x, then y may be expressed in terms of power

series in x (see section 5.2). In simple cases, the general term of the power series may be found. Progress is also possible sometimes by changing the dependent and/or the independent variable.

Independent variable. If $x = x(u)$, substitute

$$\frac{dy}{dx} = \frac{dy}{du}\bigg/\frac{dx}{du},$$

$$\frac{d^2y}{dx^2} = \frac{d}{du}\left(\frac{dy}{du}\bigg/\frac{dx}{du}\right)\frac{dx}{du} = \left(\frac{d^2y}{du^2}\frac{dx}{du}-\frac{d^2x}{du^2}\frac{dy}{du}\right)\bigg/\left(\frac{dx}{du}\right)^3.$$

Dependent variable. (i) If $y = p(x)w$, where $p(x)$ is given, then

$$Dy = p\,Dw+(Dp)\,w,$$

$$D^2y = p\,D^2w+2(Dp)\,Dw+(D^2p)\,w.$$

(ii) If $y = F(w)$, then

$$\frac{dy}{dx} = \frac{dF}{dw}\frac{dw}{dx}, \qquad \frac{d^2y}{dx^2} = \frac{d^2F}{dw^2}\left(\frac{dw}{dx}\right)^2+\frac{dF}{dw}\frac{d^2w}{dx^2}.$$

In case (i), $p(x)$ may be chosen so that, after substituting in the differential equation, the transformed equation possesses no Dw term. This is the *normal form* of the equation.

Problem 3.17 In the differential equation

$$x^2\,D^2y-10x\,Dy+24y = 5x^3,$$

change the independent variable from x to u by the transformation $x = e^u$, and find y such that $y = 2$, $Dy = 10$ when $x = 1$.

Solution. We note that

$$\frac{dy}{dx} = \frac{dy}{du}\bigg/\frac{dx}{du} = e^{-u}\frac{dy}{du},$$

$$\frac{d^2y}{dx^2} = \frac{d}{du}\left(e^{-u}\frac{dy}{du}\right)\bigg/\frac{dx}{du} = e^{-2u}\frac{d^2y}{du^2}-e^{-2u}\frac{dy}{du}.$$

Upon substitution, the equation reduces to

$$\frac{d^2y}{du^2}-11\frac{dy}{du}+24y = 5e^{3u}.$$

The complementary function is $Ae^{3u}+Be^{8u}$, and the particular integral is

$$y = \frac{1}{D^2-11D+24}\,5e^{3u} = e^{3u}\frac{1}{(D+3)^2-11(D+3)+24}\,5$$

$$= e^{3u}\frac{1}{D^2-5D}\,5 = -ue^{3u}.$$

33

Hence the general solution is
$$y = Ae^{3u} + Be^{8u} - ue^{3u} = Ax^3 + Bx^8 - x^3 \log x.$$
When $x = 1$, $y = 2 = A+B$ and $Dy = 10 = 3A+8B-1$, yielding $A = B = 1$. Thus the solution is
$$y = x^3 + x^8 - x^3 \log x. \qquad \square$$

Problem 3.18 By reducing the equation
$$D^2y - 2nx^{-1} Dy + [n^2 + n(n+1)x^{-2}] y = 0$$
to normal form, find its general solution. Deduce the solution of Bessel's equation of order $\frac{1}{2}$:
$$\frac{d^2y}{du^2} + \frac{1}{u}\frac{dy}{du} + \left(1 - \frac{1}{4u^2}\right) y = 0.$$

Solution. Transform the dependent variable from y to w using $y = p(x) w$. We have
$$p D^2w + 2(Dp) Dw + (D^2p) w - 2nx^{-1}[p Dw + (Dp) w]$$
$$+ [n^2 + n(n+1)x^{-2}] py = 0,$$
which simplifies to
$$p D^2w + 2(Dp - nx^{-1}p) Dw + [D^2p - 2nx^{-1} Dp + n^2p + n(n+1)x^{-2}p] y = 0.$$
The Dw term is eliminated when $Dp - nx^{-1}p = 0$, namely when $p = x^n$. After simplification, the equation reduces to
$$D^2w + n^2w = 0,$$
with solutions $\sin nx$, $\cos nx$. Hence
$$y = Ax^n \sin nx + Bx^n \cos nx.$$
In particular, if $n = -\frac{1}{2}$, the equation is
$$D^2y + x^{-1} Dy + (\frac{1}{4} - \frac{1}{4}x^{-2}) y = 0.$$
If we substitute $x = 2u$, the equation
$$\frac{d^2y}{du^2} + \frac{1}{u}\frac{dy}{du} + \left(1 - \frac{1}{4u^2}\right) y = 0$$
has the solution $\quad y = Mu^{-\frac{1}{2}} \sin u + Nu^{-\frac{1}{2}} \cos u.$ $\qquad \square$

Problem 3.19 A second order differential equation is given in normal form thus:
$$D^2y + k^2n^2(x) y = 0,$$
where $n^2(x)$ is a function of x. By changing the dependent variable to w, given by $y = \exp(\int x \, dx)$, where w may be expanded in inverse powers of the large parameter k, deduce approximate solutions of the equation.

Find the inequality that must be satisfied for the approximations to be realistic. Obtain these solutions for the case $n^2 = x^2$.

Solution. If $y = \exp(\int w \, dx)$, we have

$$Dy = w \exp(\int w \, dx), \qquad D^2y = w^2 \exp(\int w \, dx) + (Dw) \exp(\int w \, dx).$$

Substitution yields

$$w^2 + Dw + k^2 n^2 = 0,$$

a non-linear first order equation for w, known as a Riccati equation. If we now assume the descending series

$$w = k w_0(x) + w_1(x) + k^{-1} w_2(x) + \ldots,$$

substitution yields

$$(k w_0 + w_1 + \ldots)^2 + (k \, Dw_0 + Dw_1 + \ldots) + k^2 n^2 = 0.$$

Equating respectively the coefficients of k^2 and k to zero, we obtain

$$w_0^2 + n^2 = 0, \qquad 2 w_0 w_1 + Dw_0 = 0,$$

implying that $w_0 = \pm in$, and $w_1 = -(Dw_0)/2w_0$. Hence the approximate solutions, omitting $1/k$ terms, are

$$\begin{aligned}
y_a &= \exp\left[\int (k w_0 + w_1) \, dx\right] \\
&= \exp\left[\pm ik \int n \, dx - \int (Dw_0/2w_0) \, dx\right] \\
&= \exp\left(\pm ik \int n \, dx - \tfrac{1}{2} \log w_0\right) \\
&\propto n^{-\frac{1}{2}} \exp\left(\pm ik \int n \, dx\right).
\end{aligned}$$

When n is real and positive, the exponential factors are of unit modulus, but when n is pure imaginary, one exponential is small and the other is large in magnitude.

The approximations are valid provided that the error $D^2 y_a + k^2 n^2 y_a$ is small in magnitude compared with the term $k^2 n^2 y_a$, i.e. if

$$\left| D^2 y_a + k^2 n^2 y_a \right| \ll \left| k^2 n^2 y_a \right|,$$

$$\left| (D^2 y_a / k^2 n^2 y_a) + 1 \right| \ll 1.$$

We obtain by direct differentiation,

$$Dy_a = \pm ik n^{\frac{1}{2}} \exp(ik \int n \, dx) - \tfrac{1}{2} n^{-\frac{3}{2}} (Dn) \exp(\pm ik \int n \, dx),$$

$$\begin{aligned}
D^2 y_a = {}&- k^2 n^{\frac{3}{2}} \exp(\pm ik \int n \, dx) + \tfrac{3}{4} n^{-\frac{5}{2}} (Dn)^2 \exp(\pm ik \int n \, dx) \\
&- \tfrac{1}{2} n^{-\frac{3}{2}} (D^2 n) \exp(\pm ik \int n \, dx).
\end{aligned}$$

Upon substitution, the inequality becomes

$$\left| 1/4 k^2 n^4 \right| \left| 3(Dn)^2 - 2n D^2 n \right| \ll 1.$$

Thus k must be large and n must not have a value too near zero. The approximate solutions break down completely at and near points where n vanishes. Such points are known as *transition points*.

When $n^2 = x^2$, there is a transition point at the origin. When $x > 0$, let $n = x > 0$; when $x < 0$, let $n = -x > 0$. (This choice of branch for the square root is arbitrary). It is usual to take the lower limit of integration to be a transition point, though the upper limit must not be too near such a point, else the approximation is not valid.

For $x > 0$, $\qquad y_a = x^{-\frac{1}{2}} \exp\left(\pm ik \int_0^x x\, dx \right) = x^{-\frac{1}{2}} \exp(\pm\frac{1}{2}ikx^2);$

for $x < 0$, $\qquad y_a = (-x)^{-\frac{1}{2}} \exp\left(\pm ik \int_0^x (-x)\, dx \right)$

$$= (-x)^{-\frac{1}{2}} \exp(\mp\frac{1}{2}ikx^2). \qquad \square$$

Problem 3.20 By introducing the general change of independent variable $x = x(u)$, and then by reducing to normal form, find approximate solutions for large k of the equation

$$D^2 y + k^2 n^2(x)\, y = 0.$$

(This is an alternative approach to the previous example).

Solution. We have

$$\frac{dy}{dx} = \frac{dy}{du} \Big/ \frac{dx}{du} = \frac{y'}{x'},$$

where primes denote d/du, and

$$\frac{d^2 y}{dx^2} = \frac{d}{du}\left(\frac{y'}{x'}\right)\frac{1}{x'} = \frac{y''}{x'^2} - \frac{y'x''}{x'^3}.$$

The equation then becomes

$$y'' - \left(\frac{x''}{x'}\right)y' + k^2 n^2 x'^2 y = 0.$$

To regain normal form, let $y = p(u)\, w$. Then

$$w'' + \left(\frac{2p'}{p} - \frac{x''}{x'}\right)w' + \left(\frac{p''}{p} - \frac{x''}{x'}\frac{p'}{p} + k^2 n^2 x'^2\right)w = 0.$$

Normal form is obtained by choosing $2p'/p - x''/x'$ to vanish, namely $p = x'^{\frac{1}{2}}$. Substitution and simplification yield

$$w'' + k^2 n^2 x'^2 w = \left(\frac{3x''^2}{4x'^2} - \frac{x'''}{2x'}\right)w.$$

Provided the right-hand side bracket is small in magnitude compared with $|k^2 n^2 x'^2|$, the right-hand side is neglected. To yield approximate solutions of the required type, choose the transformation $x = x(u)$ so that

$k^2 n^2 x'^2 = -1$. With a particular branch of the square root, this implies
$$Du = ikn(x) \quad \text{or} \quad u = ik \int n(x)\, dx.$$

Solutions for w are $w = \exp(\pm u)$, and
$$y_a = x'^{\frac{1}{2}} \exp(\pm u) \propto n^{-\frac{1}{2}} \exp\left[\pm ik \int n(x)\, dx\right]$$
as before. □

Problem 3.21 In the previous two examples, find the function $n^2(x)$ such that the solutions derived there are exact and not approximate.

Solution. Various methods of procedure are possible. Here, we substitute directly into the differential equation and ensure that the residual terms vanish identically. If
$$y = n^{-\frac{1}{2}} \exp(ik \int n\, dx),$$
$$Dy = (ikn^{\frac{1}{2}} - \tfrac{1}{2}n^{-\frac{3}{2}} Dn) \exp(ik \int n\, dx),$$
$$D^2 y = \left[-k^2 n^{\frac{3}{2}} + \tfrac{3}{4}n^{-\frac{5}{2}}(Dn)^2 - \tfrac{1}{2}n^{-\frac{3}{2}} D^2 n\right] \exp(ik \int n\, dx).$$

Substitution yields
$$\tfrac{3}{4}n^{-\frac{5}{2}}(Dn)^2 - \tfrac{1}{2}n^{-\frac{3}{2}} D^2 n \equiv 0,$$
or, upon separation,
$$3Dn/n = 2D^2 n/Dn.$$

We now integrate twice in turn:
$$3\log n = 2\log(Dn) + \text{constant},$$
$$n^{\frac{3}{2}} \propto Dn,$$
$$n^{\frac{1}{2}} = Ax + B;$$
finally, $n^2 = (Ax + B)^{-4}.$ □

3.6 Physical Problems

Problem 3.22 A and B are two points separated by a distance $2b$ on a rough horizontal table, O being their midpoint. A rough particle P of mass m slides on this table along the line AB, being attached to A and B by two identical springs of natural length $a < b$ and modulus $\frac{1}{2}n^2 ma$. The coefficient of friction between the particle and the table is kn^2/g. If the particle is released from rest at a distance l from O (in the range OA), where $l > 4k$, find the position at which it next comes to rest in the range OA.

Solution. When $OP = x$, the force on P due to the spring PA is $\frac{1}{2}n^2 m(b - a - x)$ towards A, since $b - a - x$ is the extension. Similarly the force on P due to the spring PB is $\frac{1}{2}n^2 m(b - a + x)$ towards B. When motion takes place towards B, the friction $F = mg(kn^2/g)$, acting towards A. The equation of motion of P is
$$m\ddot{x} = T + F - S = mkn^2 - n^2 mx.$$

37

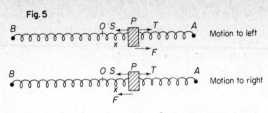

Fig. 5

Initially when $x = l$, this force equals $mn^2(k-l) < 0$ (given), so the particle will commence to move to the left, satisfying

$$\ddot{x} + n^2 x = kn^2.$$

The complementary function is $A \sin nt + B \cos nt$, and the particular integral is $x = k$. Hence generally,

$$x = k + A \sin nt + B \cos nt.$$

Initially,
$$x = l = k + B,$$
$$\dot{x} = 0 = An,$$

giving
$$x = k + (l - k) \cos nt.$$

The particle next comes to rest when $\dot{x} = 0$, namely when $t = \pi/n$, in which case,

$$x = 2k - l < 0 \qquad \text{(given)},$$

a point lying between O and B; we denote this by $-L$, where $L = l - 2k$.

For the subsequent motion towards A, F now acts to the left, so

$$\ddot{x} + n^2 x = -kn^2.$$

When $x = -L$, the acceleration is $n^2(L-k) = n^2(l-3k) > 0$, so the motion actually commences to the right. Generally,

$$x = -k + C \sin nt + D \cos nt,$$

and choosing the origin of time t at the commencement of motion towards A, we find that

$$x = -k + (3k - l) \cos nt,$$

where $x = -L$, $\dot{x} = 0$ at $t = 0$. The next position of rest occurs when $nt = \pi$, so

$$x = -k + (3k - l)(-1) = l - 4k > 0 \qquad \text{(given)},$$

a point lying in the range OA. The particle may or may not move again after this; $l > 5k$ is necessary for further movement. ☐

Problem 3.23 A long smooth straight rod is constrained to rotate in a vertical plane about one end O with constant angular velocity p. A particle P of mass m slides along the rod, and it is attached to O by a light elastic

38

string of modulus $2p^2ma$ and natural length a. At time $t = 0$, the rod is horizontal, while the particle is at rest relative to the rod with $OP = a$. Find the length $r = OP$ as a function of time t for that initial part of the motion for which the string is taut.

Solution. The radial equation of motion of the particle is

$$-T - mg \sin \theta = m(\ddot{r} - r\dot{\theta}^2) = m(\ddot{r} - rp^2)$$

in polar coordinates, the initial line being horizontal. Here, T is the tension in the string and $-mg \sin \theta$ is the component of the particle's weight directed outwards along the rod.

Now $T = 2p^2m(r - a)$ and $\theta = pt$, so

$$d^2r/dt^2 + p^2r = 2p^2a - g \sin pt.$$

Fig. 6

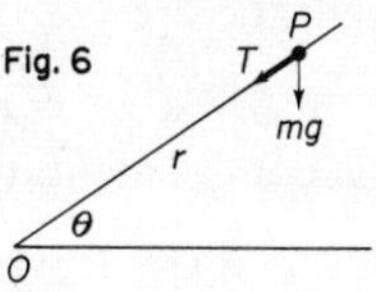

The complementary function is $A \sin pt + B \cos pt$, and the particular integral is

$$r = 2a - \mathrm{Im} \frac{ge^{ipt}}{D^2 + p^2}$$

$$= 2a - \mathrm{Im}\, ge^{ipt} \frac{1}{(D + ip)^2 + p^2} \cdot 1$$

$$= 2a - \mathrm{Im}\, ge^{ipt} \frac{1}{2ipD} \cdot 1$$

$$= 2a - \mathrm{Im}\, ge^{ipt}(t/2ip)$$

$$= 2a + (gt/2p) \cos pt.$$

The general solution is therefore

$$r = A \sin pt + B \cos pt + 2a + (gt/2p) \cos pt.$$

The initial conditions give

$$r = a = B + 2a, \qquad \dot{r} = 0 = Ap + g/2p,$$

showing that $A = -g/2p^2$, $B = -a$. Hence

$$r = -(g/2p^2) \sin pt - a \cos pt + 2a + (gt/2p) \cos pt.$$

Initially, $\dot{r} = 0$ and $\ddot{r} = ap^2$, implying that $r > a$ for the first stage of the motion. □

Problem 3.24 A vertical elastic string PQ, of natural length a and modulus mg, has a mass m attached at its lower end Q, while the upper end

39

P is constrained to oscillate vertically about a fixed point O. At time t, $OP = \alpha \sin \omega t$ and $OQ = 2a + x$, where α is small. If air resistance is given by $2kmn$ times the velocity of Q, derive the equation of motion of the mass m, where $n^2 = g/a$. If k is such that $0 < k < 1$, find the ultimate form of x as t becomes large, whatever the initial conditions at $t = 0$.

Solution. At time t, $PQ = 2a + x - \alpha \sin \omega t$, so the tension in the string is

$$T = (mg/a)(PQ - a) = (mg/a)(a + x - \alpha \sin \omega t).$$

Hence the downward equation of motion is given by

$$m\ddot{x} = mg - 2kmn\dot{x} - T = -2kmn\dot{x} - (mg/a)(x - \alpha \sin \omega t),$$

or
$$D^2 x + 2kn\, Dx + n^2 x = (g\alpha/a) \sin \omega t.$$

The roots of the auxiliary quadratic are

$$-kn \pm in \sqrt{(1 - k^2)}$$

since $1 > k^2$, given. Hence the complementary function is

$$x = A \exp\left\{\left[-kn + in \sqrt{(1 - k^2)}\right] t\right\} + B \exp\left\{\left[-kn - in \sqrt{(1 - k^2)}\right] t\right\},$$

tending to zero as $t \to \infty$. The particular integral is

$$x = \operatorname{Im} \frac{(g\alpha/a)}{D^2 + 2kn\, D + n^2} e^{i\omega t} = \operatorname{Im} \frac{(g\alpha/a)}{-\omega^2 + 2kni\omega + n^2} e^{i\omega t}$$

$$= \operatorname{Im} \frac{(g\alpha/a) e^{i\omega t}}{[(n^2 - \omega^2)^2 + 4k^2 n^2 \omega^2]^{\frac{1}{2}} e^{i\varepsilon}} = \frac{\alpha n^2 \sin(\omega t - \varepsilon)}{[(n^2 - \omega^2)^2 + 4k^2 n^2 \omega^2]^{\frac{1}{2}}},$$

where $\tan \varepsilon = 2kn\omega/(n^2 - \omega^2)$.

This is an example of a *damped oscillatory system*, caused by the factor k multiplying the derivative Dx in the differential equation. The complementary function disappears for large t leaving only the particular integral. If there is no damping,

$$x = C \sin nt + D \cos nt + \alpha n^2 \sin \omega t/(n^2 - \omega^2).$$

The solution breaks down when $\omega = n$, namely when the forcing frequency is identical with the natural resonant frequency of the system. When $\omega = n$ and $k \neq 0$, no difficulty arises; the particular integral is merely

$$x = -(\alpha/2k) \cos nt.$$

But when $\omega = n$ and $k = 0$,

$$x = \operatorname{Im} \frac{g\alpha/a}{D^2 + n^2} e^{int} = -\tfrac{1}{2} n\alpha t \cos nt,$$

an oscillatory solution with growing amplitude. $\qquad \square$

Problem 3.25 A generator producing an e.m.f. $E \cos nt$ is connected in series with a resistance R, a capacitance C and an inductance L. Examine the current I in the circuit.

Solution. There are three sources of e.m.f. in the circuit *in the direction of the current.*

(i) The given oscillatory e.m.f., $E \cos nt$.

(ii) The capacitance C with charge q (on the plate *into* which the current is indicated as entering, so $I = dq/dt \equiv Dq$). This yields an e.m.f. q/C,

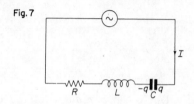

Fig. 7

acting *from* the plate with charge q *to* the plate with charge $-q$ *in the external circuit*, namely opposed to I.

(iii) The inductance L, yielding an e.m.f. $-L \, dI/dt \equiv -L \, DI$ in the direction of I.

These three sources of e.m.f. overcome the resistance of the circuit. Hence

$$E \cos nt - L \, DI - q/C = RI,$$

or, upon differentiating,

$$L D^2 I + R \, DI + I/C = -En \sin nt.$$

The complementary function is given by

$$I = Ae^{\lambda_1 t} + Be^{\lambda_2 t},$$

where

$$\lambda_1, \lambda_2 = [-R + \sqrt{(R^2 - 4L/C)}]/2L.$$

But if $R^2 = 4L/C$, we have

$$I = (A + Bt) e^{-Rt/2L},$$

tending to zero as $t \to \infty$. If $R^2 > 4L/C$, the two values of λ are real and negative, and each exponential in the complementary function tends to zero as $t \to \infty$; these are *evanescent solutions*. But if $R^2 < 4L/C$, the two values of λ are complex, and the complementary function is

$$I = e^{-Rt/2L} \left[M \cos \frac{\sqrt{(4L/C - R^2)}}{2L} t + N \sin \frac{\sqrt{(4L/C - R^2)}}{2L} t \right],$$

representing a *damped oscillatory current*. In all these cases, the complementary function tends to zero; such solutions are known as *transient currents*. The effects due to arbitrary initial conditions die out as t becomes

41

large, this phenomenon being caused by the existence of resistance in the circuit.

But when $R = 0$ (an artificial concept under normal circumstances), $\lambda_1, \lambda_2 = \pm i/\sqrt{(LC)}$, so

$$I = M \cos t/\sqrt{(LC)} + N \sin t/\sqrt{(LC)},$$

representing a natural *resonance* in the circuit that cannot decay. The angular frequency of such oscillations is given by $\omega^2 = 1/(LC)$.

The particular integral is given by

$$I = \frac{-En}{LD^2 + RD + 1/C} \sin nt = -En \operatorname{Im} \frac{1}{LD^2 + RD + 1/C} e^{int}$$

$$= -En \operatorname{Im} \frac{e^{int}}{-Ln^2 + Rin + 1/C} = E \operatorname{Im} \frac{ie^{int}}{R + i(nL - 1/nC)}.$$

Write $Z \equiv R + i(nL - 1/nC)$, the *impedance* of the circuit; R is the *resistive* component and $nL - 1/nL$ is the *reactive* component. In polar form, write $Z = |Z| e^{i\theta}$, which gives

$$I = \frac{E}{|Z|} \operatorname{Im} ie^{int - i\theta} = \frac{E}{|Z|} \cos(nt - \theta).$$

This is a *steady state* current, maintained for all time, remaining when the transient current has effectively disappeared. The angle θ shows that the current is out of phase with the e.m.f.; this phase constant vanishes only when $nL = 1/nC$.

When $R = 0$, the same theory is valid, with $\theta = \pm\frac{1}{2}\pi$, unless $n = \omega$, namely if the impressed oscillatory frequency equals the natural resonant frequency of the circuit. The particular integral must then be evaluated as follows:

$$I = -E\omega \operatorname{Im} \frac{1}{LD^2 + 1/C} e^{i\omega t}$$

$$= -E\omega \operatorname{Im} e^{i\omega t} \frac{1}{L(D + i\omega)^2 + 1/C} 1$$

$$= -E\omega \operatorname{Im} e^{i\omega t} \frac{1}{LD^2 + 2iL\omega D} 1$$

$$= -E\omega \operatorname{Im} \frac{te^{i\omega t}}{2iL\omega} = \frac{Et}{2L} \cos \omega t,$$

representing a growing oscillatory current, to which must be added the complementary function $M \cos \omega t + N \sin \omega t$. $\qquad\square$

42

EXERCISES

1. Solve the following equations:

(a) $D^2y - 8Dy + 20y = 0$,   (b) $D^2y - 4Dy + 16y = 32x^2$,

(c) $D^2y - 6Dy + 9y = 72\cosh 3x$,   (d) $D^2y - 2Dy + y = 2\cos x + 2\sin x$,

(e) $D^2y - 4Dy + 4y = 12(1+x)^2 e^{2x}$.

2. If $a + bx^2$ is one solution of the equation
$$(1+x^2)^2 D^2y - 4x(1+x^2) Dy + 2(3x^2 - 1) y = 0$$
(where a and b are to be found), find the second solution.

3. Solve the equation $D^2y + y = \cot x$.

4. Solve the differential equation $D^2y + (k^2 - 2/x^2) y = 0$, when it is given that the solution equals $e^{\pm ikx}$ times a simple polynomial in x divided by x.

5. Solve the differential equation $x^5 D^2y - xy = 1$, by changing the dependent variable to u and the independent variable to z, where $y = ux$, $x = 1/z$.

6. Expand the differential equation
$$(D + f)(D + 2f)(D + 3f) y = 0,$$
where f is a given function of x. Hence find three independent solutions of the equation
$$D^3y + 6\cot x\, D^2y + (3\cot^2 x - 8) Dy - \cot x(9 + 3\cot^2 x) y = 0.$$
[Let $(D+f)u = 0$, and find u; let $(D+2f)v = u$, and find v; let $(D+3f) y = v$, and find y.]

Answers

1. (a) $Ae^{4x} \sin 2x + Be^{4x} \cos 2x$.

   (b) $x + 2x^2 + e^{2x}[A \sin(2x\sqrt{3}) + B \cos(2x\sqrt{3})]$.

   (c) $(A + Bx)e^{3x} + 18x^2 e^{3x} + e^{-3x}$.

   (d) $(A + Bx)e^x + \cos x - \sin x$.

   (e) $(A + Bx + 6x^2 + 4x^3 + x^4)e^{2x}$.

2. $a = b = 1, x + x^3$.

3. $A \cos x + B \sin x - \sin x \log|\cot x + \operatorname{cosec} x|$.

4. $e^{\pm ikx}(1 \mp ikx)/x$.

5. $x(A^{1/x} + Be^{-1/x} - 2 - x^{-2})$.

6. $\cot^2 x \operatorname{cosec} x, \cot x \operatorname{cosec}^2 x, \operatorname{cosec}^3 x$.

43

Chapter 4

Simultaneous Differential Equations

4.1 The Solution of Simultaneous Equations The solution of n differential equations, relating the various derivatives of the n dependent variables $u, v, w, \ldots$, may be found by considering the case when $n = 2$. By differentiating and combining suitably the equations, first obtain an equation in which all the derivatives of v (say) are eliminated, but not v itself. Second, obtain an equation in which v is eliminated completely. Solve this equation for u, introducing as many arbitrary constants as necessary; the former equation then provides v in terms of this solution u and its derivatives. The *same* arbitrary constants occur in both solutions. When initial conditions are given, they should be introduced to find the values of the constants as soon as u is determined generally.

Problem 4.1 Solve the equations

$$Du + u = 3v, \qquad Dv + 2v = 2u,$$

such that $v = 0$ and $Du = 5$ when $x = 0$. No particular integral is involved.

Solution. We gather terms together thus:

$$(D+1)u = 3v, \qquad (D+2)v = 2u,$$

and we decide to eliminate v. The first equation will yield v in terms of u without further manipulation.

The variable v is eliminated by operating on the first equation by $(D+2)$ and multiplying the second equation by 3. Addition of these resulting equations yields

$$(D+2)(D+1)u + 3(D+2)v = 3(D+2)v + 6u,$$
$$(D^2 + 3D - 4)u = 0.$$

The general solution is

$$u = Ae^{-4x} + Be^{x}.$$

Initially, $v = 0$ and $Du = 5$; the first given equation then yields $u = -5$. Hence when $x = 0$,

$$u = -5 = A + B, \quad Du = 5 = -4A + B,$$

showing that $A = -2$, $B = -3$, in which case

$$u = -2e^{-4x} - 3e^{x}.$$

Finally $\qquad\qquad v = \tfrac{1}{3}(Du + u) = 2e^{-4x} - 2e^{x}.$ $\qquad\qquad$ □

Problem 4.2 The following three equations are given:

$$u+v+w = x^2, \tag{1}$$
$$(2D+3)u-v+Dw = 0, \tag{2}$$
$$u-(3D-1)v-Dw = 0. \tag{3}$$

Eliminate v and w and hence find u, v and w generally. A particular integral will be involved.

Solution. We first express v and w in terms of u and its derivatives. Operate on equation (1) by D and subtract from (2):

$$(D+3)u-(D+1)v = -2x. \tag{4}$$

Addition of equations (2) and (3) yields:

$$(2D+4)u-3Dv = 0.$$

To eliminate Dv, this equation is subtracted from 3 times equation (4):

$$(D+5)u-3v = -6x,$$
$$v = \tfrac{1}{3}(D+5)u+2x.$$

Equation (1) now yields w in terms of u:

$$w = x^2-u-v = x^2-2x-\tfrac{1}{3}(D+8)u.$$

We can eliminate v and w by substitution, say, into equation (2):

$$(2D+3)u-\tfrac{1}{3}(D+5)u-2x+2x-2-\tfrac{1}{3}D(D+8)u = 0,$$
$$(D^2+3D-4)u = -6.$$

The solution is

$$u = Ae^x+Be^{-4x}+\tfrac{3}{2}.$$

Finally,
$$v = \tfrac{1}{3}(D+5)u+2x = 2Ae^x+Be^{-4x}/3+2x+\tfrac{5}{2},$$
$$w = -(u+v)+x^2 = -3Ae^x-4Be^{-4x}/3+x^2-2x-4. \qquad \square$$

Sometimes, two given equations bear a special relation the one to the other. If the second equation is obtained from the first by changing u and v to v and u respectively, addition and subtraction yield two non-simultaneous equations. Again, if u and v in the first are replaced by v and $-u$ in the second, multiplication of the second equation by i and then addition to, and subtraction from, the first equation yield two non-simultaneous equations. When the coefficients are real, only addition is needed. This latter type occurs in oscillatory problems with two degrees of freedom.

Problem 4.3 Solve the equations

$$D^2u+2Dv+u = \sin x, \qquad D^2v+2Du+v = \cos x,$$

such that $u=v=Du=Dv=0$ when $x=0$.

Solution. Addition and subtraction yield

$$D^2(u+v)+2D(u+v)+(u+v) = \sin x + \cos x,$$
$$D^2(u-v)-2D(u-v)+(u-v) = \sin x - \cos x.$$

Writing $u+v = p$, $u-v = q$, where $p = q = Dp = Dq = 0$ when $x = 0$, we have

$$(D^2+2D+1)p = (\mathrm{Im}+\mathrm{Re})\,e^{ix},$$
$$(D^2-2D+1)q = (\mathrm{Im}-\mathrm{Re})\,e^{ix}.$$

The complementary functions are $(A+Bx)\,e^{-x}$ and $(M+Nx)\,e^{x}$ respectively, and the particular integrals are

$$p = (\mathrm{Im}+\mathrm{Re})\frac{1}{(D+1)^2}e^{ix} = (\mathrm{Im}+\mathrm{Re})\frac{e^{ix}}{(1+i)^2}$$
$$= \tfrac{1}{2}(\sin x - \cos x),$$
$$q = (\mathrm{Im}-\mathrm{Re})\frac{1}{(D-1)^2}e^{ix} = (\mathrm{Im}-\mathrm{Re})\frac{e^{ix}}{(i-1)^2}$$
$$= \tfrac{1}{2}(\sin x + \cos x).$$

Generally, the solutions are

$$p = (A+Bx)\,e^{-x}+\tfrac{1}{2}(\sin x - \cos x),$$
$$q = (M+Nx)\,e^{x}+\tfrac{1}{2}(\sin x + \cos x).$$

When $x = 0$,

$$p = 0 = A-\tfrac{1}{2}, \qquad Dp = 0 = -A+B+\tfrac{1}{2},$$
$$q = 0 = M+\tfrac{1}{2}, \qquad Dq = 0 = M+N+\tfrac{1}{2},$$

yielding $A = \tfrac{1}{2}$, $B = 0$, $M = -\tfrac{1}{2}$, $N = 0$. Hence

$$p = \tfrac{1}{2}(e^{-x}+\sin x - \cos x),$$
$$q = \tfrac{1}{2}(-e^{x}+\sin x + \cos x),$$
$$u = \tfrac{1}{2}(p+q) = \tfrac{1}{2}(\sin x - \sinh x),$$
$$v = \tfrac{1}{2}(p-q) = \tfrac{1}{2}(\cosh x - \cos x). \qquad \square$$

Problem 4.4 Solve the equations

$$D^2u+5Dv-4u = 18\sin 2x, \qquad D^2v-5Du-4v = 0,$$

such that $u = v = Dv = 0$, $Du = 6$ when $x = 0$.

Solution. Let $z = u+iv$, where $z = 0$ and $Dx = 6$ when $x = 0$. Multiply the second equation by i and add to the first:

$$D^2z-5iDz-4z = 18\sin 2x.$$

The roots of the auxiliary quadratic $\lambda^2-5i\lambda-4 = 0$ are $4i$ and i, so the complementary function is $Ae^{4ix}+Be^{ix}$.

To find the particular integral, we cannot express the right-hand side as Im $18e^{2ix}$, since the coefficients on the left-hand side are complex; the 'i's in the usual process would become intermixed with the 'i's in the coefficients, and the final calculation of the imaginary part would yield the wrong answer. Thus we write directly

$$(D^2 - 5iD - 4)z = 9(e^{2ix} - e^{-2ix})/i,$$

and
$$z = \frac{9}{i}\left(\frac{1}{D^2 - 5iD - 4}e^{2ix} - \frac{1}{D^2 - 5iD - 4}e^{-2ix}\right)$$

$$= \frac{9}{i}\left(\frac{e^{2ix}}{2} - \frac{e^{-2ix}}{-18}\right) \qquad \text{(replacing } D \text{ by } 2i \text{ and } -2i \text{ respectively)},$$

$$z = Ae^{4ix} + Be^{ix} - 9ie^{2ix}/2 - ie^{-2ix}/2.$$

Initially,

$$0 = z = A + B - 9i/2 - i/2, \qquad 6 = Dz = 4iA + iB + 9 - 1,$$

yielding $A = -i$, $B = 6i$. Finally,

$$u = \text{Re } z = \text{Re}(-ie^{4ix} + 6ie^{ix} - 9ie^{2ix}/2 - ie^{-2ix}/2)$$

$$= \sin 4x - 6 \sin x + 4 \sin 2x,$$

$$v = \text{Im } z = -\cos 4x + 6 \cos x - 5 \cos 2x. \qquad \square$$

4.2 Matrix Methods The following methods, illustrating two simultaneous equations, are immediately applicable to n equations.

Problem 4.5   Solve the system of equations

$$Du = -13u + 4v + 1,$$
$$Dv = -42u + 13v + x,$$

using matrix methods.

Solution. In matrix notation, we write

$$D\begin{pmatrix} u \\ v \end{pmatrix} = \begin{pmatrix} -13 & 4 \\ -42 & 13 \end{pmatrix}\begin{pmatrix} u \\ v \end{pmatrix} + \begin{pmatrix} 1 \\ x \end{pmatrix},$$

or
$$D\mathbf{u} = \mathbf{A}\mathbf{u} + \mathbf{f},$$

the matrix symbols standing for the explicit matrices respectively.

Introduce the transformation $\mathbf{u} = \mathbf{T}\mathbf{p}$, where $\mathbf{p}$ is a new dependent column consisting of the elements p and q. Then $D\mathbf{u} = \mathbf{T}\,D\mathbf{p}$ provided $\mathbf{T}$ is a constant square matrix, and the equations become

$$\mathbf{T}\,D\mathbf{p} = \mathbf{A}\,\mathbf{T}\mathbf{p} + \mathbf{f},$$
$$D\mathbf{p} = \mathbf{T}^{-1}\mathbf{A}\,\mathbf{T}\mathbf{p} + \mathbf{T}^{-1}\mathbf{f}, \qquad (1)$$

where $\mathbf{T}$ must be non-singular.

We choose $\mathbf{T}$ such that $\mathbf{A}$ is diagonalized. The characteristic equation of $\mathbf{A}$ is given by

$$\begin{vmatrix} -13-\lambda & 4 \\ -42 & 13-\lambda \end{vmatrix} = 0,$$

namely, $\lambda^2 - 1 = 0$ with roots $\lambda = 1, -1$. The corresponding characteristic vectors that ensure that $\mathbf{Ak} = \lambda\mathbf{k}$ are

$$\binom{2}{7} \quad \text{and} \quad \binom{1}{3}.$$

If we write $\qquad \mathbf{T} = \begin{pmatrix} 2 & 1 \\ 7 & 3 \end{pmatrix}, \qquad \mathbf{T}^{-1} = \begin{pmatrix} -3 & 1 \\ 7 & -2 \end{pmatrix}$

standard matrix theory yields

$$\mathbf{T}^{-1}\mathbf{AT} = \begin{pmatrix} \lambda_1 & 0 \\ 0 & \lambda_2 \end{pmatrix} = \begin{pmatrix} 1 & 0 \\ 0 & -1 \end{pmatrix}.$$

Equation (1) becomes

$$D\mathbf{p} = \begin{pmatrix} 1 & 0 \\ 0 & -1 \end{pmatrix}\mathbf{p} + \begin{pmatrix} -3 & 1 \\ 7 & -2 \end{pmatrix}\binom{1}{x},$$

or $\qquad Dp = p - 3 + x, \qquad Dq = -q + 7 - 2x.$

The solutions of these two non-simultaneous equations are

$$p = Ae^x + 3 - \frac{1}{1-D}x = Ae^x + 3 - (x+1) = Ae^x + 2 - x,$$

$$q = Be^{-x} + 7 - \frac{1}{1+D}2x = Be^{-x} + 7 - (2x-2) = Be^{-x} + 9 - 2x.$$

Finally, $\qquad \binom{u}{v} = \mathbf{Tp} = \begin{pmatrix} 2 & 1 \\ 7 & 3 \end{pmatrix}\binom{Ae^x+2-x}{Be^{-x}+9-2x},$

i.e. $\qquad u = 2Ae^x + Be^{-x} + 13 - 4x, \qquad v = 7Ae^x + 3Be^{-x} + 41 - 13x.$ □

Problem 4.6 Solve the system of equations

$$Du = -8u + 4v + 3e^x,$$
$$Dv = -9u + 4v + e^{-x},$$

using matrix methods, where the system matrix cannot be diagonalized.

Solution. In matrix notation, we write

$$D\binom{u}{v} = \begin{pmatrix} -8 & 4 \\ -9 & 4 \end{pmatrix}\binom{u}{v} + \binom{3e^x}{e^{-x}},$$

$$D\mathbf{u} = \mathbf{Au} + \mathbf{f}.$$

48

If $\mathbf{T}$ is a constant non-singular transformation matrix, let $\mathbf{u} = \mathbf{Tp}$ with $D\mathbf{u} = \mathbf{T}\,D\mathbf{p}$; hence

$$\mathbf{T}\,D\mathbf{p} = \mathbf{ATp} + \mathbf{f},$$
$$D\mathbf{p} = \mathbf{T}^{-1}\mathbf{ATp} + \mathbf{T}^{-1}\mathbf{f}. \tag{1}$$

The characteristic equation of $\mathbf{A}$ is given by

$$\begin{vmatrix} -8-\lambda & 4 \\ -9 & 4-\lambda \end{vmatrix} = 0,$$

with roots -2 and -2. One characteristic vector has elements 2, 3, but there exists no other, implying that the matrix $\mathbf{A}$ cannot be diagonalized. The simplest transformed form for $\mathbf{A}$ must be taken to be

$$\begin{pmatrix} -2 & 1 \\ 0 & -2 \end{pmatrix},$$

and $\mathbf{T}$ must then be written

$$\mathbf{T} = \begin{pmatrix} 2 & a \\ 3 & b \end{pmatrix}.$$

Standard matrix theory gives

$$\mathbf{AT} = \mathbf{T}\begin{pmatrix} -2 & 1 \\ 0 & -2 \end{pmatrix},$$

$$\begin{pmatrix} -8 & 4 \\ -9 & 4 \end{pmatrix}\begin{pmatrix} 2 & a \\ 3 & b \end{pmatrix} = \begin{pmatrix} 2 & a \\ 3 & b \end{pmatrix}\begin{pmatrix} -2 & 1 \\ 0 & -2 \end{pmatrix}.$$

Multiplication yields simple equations with solutions $a = -3$, $b = -4$. Hence

$$\mathbf{T} = \begin{pmatrix} 2 & -3 \\ 3 & -4 \end{pmatrix} \quad \text{and} \quad \mathbf{T}^{-1} = \begin{pmatrix} -4 & 3 \\ -3 & 2 \end{pmatrix}.$$

Equation (1) becomes

$$D\mathbf{p} = \begin{pmatrix} -2 & 1 \\ 0 & -2 \end{pmatrix}\mathbf{p} + \begin{pmatrix} -4 & 3 \\ -3 & 2 \end{pmatrix}\mathbf{f}.$$

These equations are not non-simultaneous; explicitly we have

$$Dp = -2p + q - 12e^x + 3e^{-x},$$
$$Dq = -2q - 9e^x + 2e^{-x}.$$

The second equation is solved for q and then the first for p. We have

$$q = Ae^{-2x} - \frac{9}{D+2}e^x + \frac{2}{D+2}e^{-x} = Ae^{-2x} - 3e^x + 2e^{-x}.$$

The equation for p then becomes

$$Dp+2p = q-12e^x+3e^{-x} = Ae^{-2x}-15e^x+5e^{-x},$$

with
$$p = Be^{-2x}+\frac{A}{D+2}e^{-2x}-\frac{15}{D+2}e^x+\frac{5}{D+2}e^{-x}$$

$$= Be^{-2x}+Axe^{-2x}-5e^x+5e^{-x}.$$

Finally,
$$\mathbf{u} = \mathbf{Tp} = \begin{pmatrix} 2 & -3 \\ 3 & -4 \end{pmatrix}\begin{pmatrix} Axe^{-2x}+Be^{-2x}-5e^x+5e^{-x} \\ Ae^{-2x}-3e^x+2e^{-x} \end{pmatrix},$$

and
$$u = A(2x-3)e^{-2x}+2Be^{-2x}-e^x+4e^{-x},$$

$$v = A(3x-4)e^{-2x}+3Be^{-2x}-3e^x+7e^{-x}. \qquad \square$$

4.3 Phase-plane Analysis   Two simultaneous equations in $u(x)$ and $v(x)$, containing only the first order derivatives of u and v, may be formally solved to yield Du and Dv as functions of u, v and perhaps x. We write

$$Du = U(u,v), \qquad Dv = V(u,v).$$

When the independent variable x does not occur explicitly on the right-hand sides, the system is known as *autonomous*. (The variable t is more often used for x in this context, and we shall henceforth use t as the independent variable throughout this section.) The Cartesian plane Ou, v, known as the *phase-plane*, may be used to represent any solution $u = u(t), v = v(t)$ as a curve defined parametrically; such curves are known as *trajectories*. When $\partial U/\partial u$, $\partial U/\partial v$, $\partial V/\partial u$, $\partial V/\partial v$ are continuous, and if $U = V = 0$ at the point (u_0, v_0), then this point is known as an *equilibrium point*. One particular solution is obviously the constant solution $u = u_0$, $v = v_0$, representing a 'point trajectory' at the equilibrium point. The equilibrium point P is *asymptotically stable* if trajectories approach P as a limit as $t \to \infty$. The point P is *unstable* if trajectories pass off to infinity as $t \to \infty$. *Neutral stability* occurs when the trajectories do not approach P, but remain at a bounded distance from P for large t.

When U and V are linear functions of u and v, the investigation of stability is straight forward, but when the functions are not linear, a Taylor expansion about P yields linear terms, quadratic terms, and so on. If the investigation of the linear terms is definitive (cases I to V below), the question of the stability of the equilibrium is thereby settled, but if not, then the quadratic terms must also be taken into account. This latter question is not treated in this text.

Problem 4.7   Examine the trajectories in the phase-plane when $U(u,v)$ and $V(u,v)$ are linear functions of u and v, the origin being the equilibrium point. Give examples.

Solution. The given equations are

$$Du = au+bv, \quad Dv = cu+dv, \quad (a,b,c,d \text{ real}).$$

In matrix notation, we write

$$\mathbf{Du} \equiv D\begin{pmatrix} u \\ v \end{pmatrix} = \begin{pmatrix} a & b \\ c & d \end{pmatrix}\begin{pmatrix} u \\ v \end{pmatrix}.$$

If we introduce the transformation $\mathbf{u} = \mathbf{Tp}$, $\mathbf{T}$ being constant,

$$D\mathbf{p} = \mathbf{T}^{-1}\mathbf{ATp}.$$

Let the characteristic roots of $\mathbf{A}$ be λ_1 and λ_2; if these are distinct, let $\mathbf{T}$ diagonalize $\mathbf{A}$, yielding

$$Dp = \lambda_1 p, \quad Dq = \lambda_2 q, \quad \mathbf{u} = \mathbf{T}\begin{pmatrix} Ae^{\lambda_1 t} \\ Be^{\lambda_2 t} \end{pmatrix}.$$

But if the two roots are not distinct, let $\mathbf{T}$ be such that

$$D\mathbf{p} = \begin{pmatrix} \lambda & 1 \\ 0 & \lambda \end{pmatrix}\mathbf{p},$$

possessing the solution $\quad \mathbf{u} = \mathbf{T}\begin{pmatrix} Ae^{\lambda t}+Bte^{\lambda t} \\ Be^{\lambda t} \end{pmatrix}.$

The characteristic roots may take the following forms:

	Character	λ_1	λ_2	Nature of P
I	real	$+$	$-$	saddle-point, unstable
II	real	$+$	$+$	node, unstable
III	real	$-$	$-$	node, asymp. stable
IV	complex $\mu \pm iv$	$\mu>0$	$\mu>0$	focus, unstable
V	complex $\mu \pm iv$	$\mu<0$	$\mu<0$	focus, asymp. stable
VI	real	$+$	0	
VII	real	$-$	0	
VIII	real, $\lambda_1 = \lambda_2$	$+$	$+$	
IX	real, $\lambda_1 = \lambda_2$	$-$	$-$	
X	pure imaginary	$+iv$	$-iv$	centre, neutral
XI		0	0	

Case I. If A and B do not vanish, clearly the point $\mathbf{u}$ can never attain the equilibrium point (the origin) for any value of t; as $t \to \infty$ only one of the exponential functions involved in the solution tends respectively to zero. Exceptionally, if one of A or B vanishes, u/v will equal a constant, so the trajectory will be a straight line. Otherwise the trajectories will pass from infinity to infinity within the four sectors formed by these lines. Such an equilibrium point is known as a *saddle-point*, and is unstable.

51

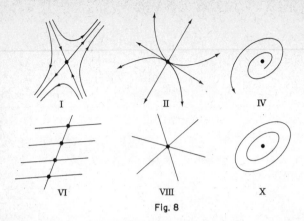

Fig. 8

Case II. The solutions u and v tend to infinity as $t \to \infty$, but approach the origin as $t \to -\infty$. Hence all trajectories emerge from the equilibrium point, and curve out to infinity with increasing time. This configuration is known as an *unstable node*.

Case III. This is Case II reversed in time; all trajectories converge to the origin, which is known as a *stable node*.

Case IV. $|u|$ and $|v| \to \infty$ as $t \to \infty$, but tend to zero as $t \to -\infty$. The trajectories spiral outwards (a feature produced by the terms $e^{\pm ivt}$) from the origin, yielding an *unstable focus*.

Case V. This is Case IV reversed in time; all trajectories spiral into the origin, yielding a *stable focus*.

The remaining cases depend critically upon the values of a, b, c, d.

Case VI. The solution will be

$$u = T_{11} A e^{\lambda t} + T_{12} B, \qquad v = T_{21} A e^{\lambda t} + T_{22} B,$$

consisting of a series of parallel straight lines. The vanishing of one root demands

$$\begin{vmatrix} a & b \\ c & d \end{vmatrix} = 0.$$

Hence, if (u_0, v_0) is an equilibrium point, so is (ku_0, kv_0), implying that there is a straight line of equilibrium points passing through the origin. Each trajectory passes from an equilibrium point out to infinity.

Case VII. This is Case VI reversed in time; each straight line trajectory terminates on the line of equilibrium points.

52

Case VIII. If $a = d$, $b = c = 0$, the trajectories form a family of straight lines emerging from the origin, which is an unstable equilibrium point. If $bc \neq 0$, the trajectories form a series of curves emerging from the origin.

Case IX. This is case VIII reversed in time; the origin is stable.

Case X. The trajectories tend neither to the origin nor to infinity. In fact, they form ellipses, with the origin as centre; this is a case of neutral stability.

Case XI. If $a = b = c = d = 0$, all points are stable equilibrium points. Otherwise, all points on the line $au + bv = 0$ are equilibrium points. All trajectories are lines parallel to this line.

These may be illustrated by the following examples; the reader may check these solutions, and sketch the simpler families of trajectories.

Case I.
$$Du = 3u - 2v, \qquad Dv = 4u - 3v; \qquad (\lambda = \pm 1),$$
$$u = Ae^t + Be^{-t}, \qquad v = Ae^t + 2e^{-t},$$
$$(v - u)(2u - v) = AB.$$

Case II.
$$Du = v, \qquad Dv = -2u + 3v; \qquad (\lambda = 1, 2),$$
$$u = Ae^t + Be^{2t}, \qquad v = Ae^t + 2Be^{2t},$$
$$(2u - v)^2 = A^2(v - u)/B.$$

Case III.
$$Du = -v, \qquad Dv = 2u - 3v; \qquad (\lambda = -1, -2),$$
$$u = Ae^{-t} + Be^{-2t}, \qquad v = Ae^{-t} + 2Be^{-2t},$$
the (u, v)-trajectories being the same as in case II.

Case IV.
$$Du = 2u + 2v, \qquad Dv = -u; \qquad (\lambda = 1 \pm i),$$
$$u = e^t(2A \cos t + 2B \sin t), \qquad v = e^t[(B - A)\cos t - (A + B)\sin t],$$
$$2 \tan^{-1} \frac{(B - A)u - 2Av}{(A + B)u + 2Bv} = \log \frac{u^2 + 2uv + 2v^2}{2(A^2 + B^2)}.$$

Case V.
$$Du = -2u - 2v, \qquad Dv = u; \qquad (\lambda = -1 \pm i),$$
$$u = e^{-t}(2A \cos t - 2B \sin t), \qquad v = e^{-t}[(B - A)\cos t + (A + B)\sin t],$$
the (u, v)-trajectories being the same as in case IV.

Case VI.
$$Du = 2u - v, \qquad Dv = 2u - v; \qquad (\lambda = 1, 0),$$
$$u = Ae^t + B, \qquad v = Ae^t + 2B,$$
$$u - v + B = 0.$$

53

Case VII. $\qquad Du = -2u+v, \quad Dv = -2u+v; \qquad (\lambda = -1,0),$

$$u = Ae^{-t}+B, \qquad v = Ae^{-t}+2B,$$

the (u,v)-trajectories being the same as in case VI.

Case VIIIa. $\qquad Du = u, \quad Dv = v; \qquad (\lambda = 1,1),$

$$u = Ae^t, \quad v = Be^t,$$
$$v = Bu/A.$$

Case VIIIb. $\qquad Du = v, \quad Dv = -u+2v; \qquad (\lambda = 1,1),$

$$u = [A+B(1+t)]\,e^t, \quad v = [A+B(2+t)]\,e^t,$$
$$(v-u)\{A/B+2+\log[(v-u)/B]\} = v.$$

Case IXa. $\qquad Du = -u, \quad Dv = -v, \qquad (\lambda = -1,-1),$

$$u = Ae^{-t}, \qquad v = Be^{-t},$$

the (u,v)-trajectories being the same as in case VIIIa.

Case IXb. $\qquad Du = -v, \quad Dv = u-2v; \qquad (\lambda = -1,-1),$

$$u = [A+B(1-t)]\,e^{-t}, \quad v = [A+B(2-t)]\,e^{-t},$$

the (u,v)-trajectories being the same as in case VIIIb.

Case X. $\qquad Du = u+2v, \quad Dv = -u-v; \qquad (\lambda = \pm i),$

$$u = 2A\cos t+2B\sin t, \quad v = (B-A)\cos t-(A+B)\sin t,$$
$$u^2+2uv+2v^2 = 2(A^2+B^2).$$

Case XIa. $\qquad Du = 0, \quad Dv = 0; \qquad (\lambda = 0,0),$

$$u = A, \quad v = B.$$

Case XIb. $\qquad Du = -u+v, \quad Dv = -u+v; \qquad (\lambda = 0,0),$

$$u = At+B, \qquad v = At+A+B,$$
$$u-v+A = 0.$$

4.4 Normal Modes of Oscillation If two (or more) variables u and v satisfy simultaneous differential equations with oscillatory solutions, a normal mode is a particular solution in which u and v have the same period and phase constant. There will be several independent normal modes corresponding to the same frequency. The general solution consists of a linear combination of the normal modes, the constants in the linear combination being chosen so as to satisfy given initial conditions.

Problem 4.8 A light string of length $7a$ has two particles of equal mass m attached to one end and to a point distance $4a$ from that end. The string

54

hangs freely from the other end under gravity. If the system performs small oscillations in which the particles move approximately horizontally in the same vertical plane, show that the equations for the displacements u and v of the upper and lower particles are

$$(12D^2 + 11n^2)u = 3n^2 v, \qquad (4D^2 + n^2)v = n^2 u$$

where $n^2 = g/a$. Prove that u and v are each the sum of two oscillations of periods $2\pi/n$ and $2\pi(\sqrt{6})/n$. Find u and v in terms of t, given that, at $t = 0$, $Du = Dv = 0$, $u = b$, $v = 0$.

Solution. Denote the two particles by A and B respectively, with B below

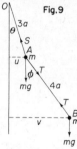

Fig.9

A. Let S and T denote the tensions in the portions of string OA and AB respectively, these portions making the angles θ and ϕ with the downward vertical.

The horizontal equations of motion are:

$$m D^2 u = T \sin\phi - S \sin\theta, \qquad m D^2 v = -T \sin\phi,$$

while vertical equilibrium demands

$$T = mg, \qquad S = T + mg = 2mg.$$

Geometrically, $\sin\phi = (v-u)/4a$, $\sin\theta = u/3a$. Upon substitution

$$(12D^2 + 11n^2)u = 3n^2 v, \qquad (4D^2 + n^2)v = n^2 u.$$

The form of the equations suggests that we may take solutions in the form

$$u = U \cos pt \quad (\text{or } \sin pt), \qquad v = V \cos pt \quad (\text{or } \sin pt),$$

yielding upon substitution,

$$(-12p^2 + 11n^2)U - 3n^2 V = 0, \qquad (-4p^2 + n^2)V - n^2 U = 0.$$

To eliminate U and V, we write

$$\begin{vmatrix} -12p^2 + 11n^2 & -3n^2 \\ -n^2 & -4p^2 + n^2 \end{vmatrix} = 0,$$

$$48p^4 - 56n^2 p^2 + 8n^4 = 0$$

with solutions $p = n$ and $n/\sqrt{6}$ (only positive values are needed).

When $p^2 = n^2$, we need $U + 3V = 0$, so we may take $U = 3$, $V = -1$. Hence the normal modes are

$$u_1 = 3 \cos nt, \qquad v_1 = -\cos nt.$$
$$u_2 = 3 \sin nt, \qquad v_2 = -\sin nt,$$

with period $2\pi/n$.

When $p^2 = n^2/6$, we need $9U - 3V = 0$, so we may take $U = 1$, $V = 3$. Hence the normal modes are

$$u_3 = \cos(nt/\sqrt{6}), \qquad v_3 = 3 \cos(nt/\sqrt{6});$$
$$u_4 = \sin(nt/\sqrt{6}), \qquad v_4 = 3 \sin(nt/\sqrt{6}),$$

with period $2\pi(\sqrt{6})/n$.

The general solution is

$$u = 3A \cos nt + 3B \sin nt + C \cos(nt/\sqrt{6}) + D \sin(nt/\sqrt{6}),$$
$$v = -A \cos nt - B \sin nt + 3C \cos(nt/\sqrt{6}) + 3D \sin(nt/\sqrt{6}).$$

When $t = 0$, we are given

$$u = b = 3A + C,$$
$$v = 0 = -A + 3C,$$
$$Du = 0 \propto 3B + D/\sqrt{6},$$
$$Dv = 0 \propto -B + 3D/\sqrt{6}.$$

Hence $A = 3b/10$, $B = 0$, $C = b/10$, $D = 0$. $\qquad\square$

The use of matrix theory greatly facilitates the calculation of normal modes, as the following example shows.

Problem 4.9 Three particles, each of mass m, are attached at equal intervals a to a light string of length $4a$ which is stretched to a tension T and fixed at the ends to a smooth horizontal table. Write down the equations of motion for small horizontal transverse vibrations of the system, and find the normal modes of oscillation.

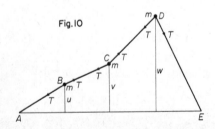

Fig.10

Solution. Let the string be represented by $ABCDE$, where A and E are fixed in a horizontal line. B, C, and D denote the particles at distances u, v, w from the line AE.

Resolving the tensions perpendicular to AE, we find that the equations of motion are

$$m\,D^2u = -Tu/a + T(v-u)/a,$$
$$m\,D^2v = -T(v-u)/a + T(w-v)/a,$$
$$m\,D^2w = -T(w-v)/(a - Tw/a,$$

or

$$D^2\begin{pmatrix} u \\ v \\ w \end{pmatrix} = n^2 \begin{pmatrix} -2 & 1 & 0 \\ 1 & -2 & 1 \\ 0 & 1 & -2 \end{pmatrix}\begin{pmatrix} u \\ v \\ w \end{pmatrix}$$

where $n^2 = T/ma$. More simply, we may write

$$D^2\mathbf{u} = n^2\,\mathbf{A}\mathbf{u}.$$

Introduce the transformation $\mathbf{u} = \mathbf{T}\mathbf{f}$, where $\mathbf{T}$ is a constant matrix. Then

$$D^2\mathbf{f} = n^2\mathbf{T}^{-1}\,\mathbf{A}\mathbf{T}\mathbf{f};$$

in particular, we may diagonalize $\mathbf{A}$. The characteristic roots of $\mathbf{A}$ are the solutions of $|\mathbf{A} - \lambda\mathbf{I}| = 0$, i.e.

$$\lambda^3 + 6\lambda^2 + 10\lambda + 4 = 0,$$
$$(\lambda+2)(\lambda-2-\sqrt{2})(\lambda-2+\sqrt{2}) = 0,$$

so $\lambda = -2,\ -2-\sqrt{2},\ -2+\sqrt{2}$. The corresponding characteristic vectors may be checked to be $(1, 0, 1)$, $(1, -\sqrt{2}, 1)$, $(1, \sqrt{2}, 1)$ respectively. Hence we take

$$\mathbf{T} = \begin{pmatrix} 1 & 1 & 1 \\ 0 & -\sqrt{2} & \sqrt{2} \\ -1 & 1 & 1 \end{pmatrix}.$$

Thus

$$D^2\mathbf{f} = n^2 \begin{pmatrix} -2 & 0 & 0 \\ 0 & -2-\sqrt{2} & 0 \\ 0 & 0 & -2+\sqrt{2} \end{pmatrix}\mathbf{f},$$

$$D^2 f + 2n^2 f = 0,$$
$$D^2 g + (2+\sqrt{2})\,n^2 g = 0,$$
$$D^2 h + (2-\sqrt{2})\,n^2 h = 0.$$

Solutions are
$$f = F_1 \cos\left[(\sqrt{2})\,nt\right] + F_2 \sin\left[(\sqrt{2})\,nt\right],$$
$$g = G_1 \cos\left[\sqrt{(2+\sqrt{2})}\,nt\right] + G_2 \sin\left[\sqrt{(2+\sqrt{2})}\,nt\right],$$
$$h = H_1 \cos\left[\sqrt{(2-\sqrt{2})}\,nt\right] + H_2 \sin\left[\sqrt{(2-\sqrt{2})}\,nt\right].$$

Normal modes are obtained by choosing five of the constants to vanish. Then

$$\begin{pmatrix} u \\ v \\ w \end{pmatrix} = \begin{pmatrix} 1 & 1 & 1 \\ 0 & -\sqrt{2} & \sqrt{2} \\ -1 & 1 & 1 \end{pmatrix} \begin{pmatrix} \cos\left[(\sqrt{2})\,nt\right] \\ 0 \\ 0 \end{pmatrix}$$

yields

$$u = \cos\left[(\sqrt{2})\,nt\right], \quad v = 0, \quad w = -\cos\left[(\sqrt{2})\,nt\right],$$

with a similar solution involving sines. The other normal modes involving cosines are

$$(u, v, w) = (1, -\sqrt{2}; 1)\cos\left[\sqrt{(2+\sqrt{2})}\,nt\right],$$
$$(u, v, w) = (1, \sqrt{2}, 1)\cos\left[(2-\sqrt{2})\,nt\right]. \qquad \square$$

The appropriate differential equations may often be obtained by the use of Lagrange's equations. If $L = T - V$, where T and V denote the kinetic and potential energies of an oscillatory system expressed to the second order in the small quantities u, v, $Du \equiv \dot{u}$, $Dv \equiv \dot{v}$, where u and v denote displacements from equilibrium values, then

$$\frac{d}{dt}\left(\frac{\partial L}{\partial \dot{u}}\right) - \frac{\partial L}{\partial u} = 0, \qquad \frac{d}{dt}\left(\frac{\partial L}{\partial \dot{v}}\right) - \frac{\partial L}{\partial v} = 0.$$

Problem 4.10 A uniform rod AB, of mass $3m$ and length $3a$, is smoothly hinged to a fixed support at A. A light inextensible string of length a is attached at B and carries a particle of mass m at its other end C. The system executes small oscillations in a vertical plane about the position of stable equilibrium. Find the kinetic and potential energies correct to the

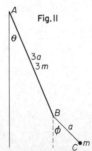

Fig. II

second order in the angles θ and ϕ which AB and BC make respectively with the downward vertical. Hence obtain the normal modes of vibration.

Solution. The moment of inertia of the rod AB is $(4/3)\,3m(3a/2)^2 = 9ma^2$, so its kinetic energy is $(9/2)\,ma^2\,\dot\theta^2$. The coordinates of C are $(3a\cos\theta + a\cos\phi)$ and $(3a\sin\theta + a\sin\phi)$, so its velocity components are $-(3a\sin\theta\,\dot\theta + a\sin\phi\,\dot\phi)$ and $(3a\cos\theta\,\dot\theta + \cos\phi\,\dot\phi)$, and its kinetic energy is $\frac{1}{2}ma^2[(3\sin\theta\,\dot\theta + \sin\phi\,\dot\phi)^2 + (3\cos\theta\,\dot\theta + \cos\phi\,\dot\phi)^2] \equiv \frac{1}{2}ma^2(9\dot\theta^2 + 6\dot\theta\dot\phi + 6\dot\phi)^2$ to the second order, since $\sin\theta = \theta$, $\sin\phi = \phi$. Hence, totally,

$$T = ma^2(9\dot\theta^2 + 3\dot\theta\dot\phi + \tfrac{1}{2}\dot\phi^2).$$

The potential energy of the system is obviously

$$V = -3mg\,[(3a/2)\cos\theta] - mg(3a\cos\theta + a\cos\phi)$$

with respect to the horizontal level through A, or

$$V = \tfrac{1}{4}mga(15\theta^2 + 2\phi^2)$$

neglecting constant terms, where $\cos\theta = 1 - \tfrac{1}{2}\theta^2$.

The Lagrangian function L is

$$L \equiv T - V = ma^2(9\dot\theta^2 + 3\dot\theta\dot\phi + \tfrac{1}{2}\dot\phi^2) - \tfrac{1}{4}mga(15\theta^2 + 2\phi^2).$$

Hence the Lagrange equations of motion are

$$\frac{d}{dt}\left(\frac{\partial L}{\partial \dot\theta}\right) - \frac{\partial L}{\partial \theta} = \frac{d}{dt}\left[ma^2(18\dot\theta + 3\dot\phi)\right] + \tfrac{15}{2}mga\,\theta$$

$$= ma^2(18\ddot\theta + 3\ddot\phi) + 15mga\,\theta/2 = 0,$$

$$\frac{d}{dt}\left(\frac{\partial L}{\partial \dot\phi}\right) - \frac{\partial L}{\partial \phi} = \frac{d}{dt}\left[ma^2(3\dot\theta + \dot\phi)\right] + mga\,\phi$$

$$= ma^2(3\ddot\theta + \ddot\phi) + mga\,\phi = 0.$$

If $\theta = P\cos nt$, $\phi = Q\cos nt$ (and similarly for $\sin nt$),

$$ma^2(18Pn^2 + 3Qn^2) - 15mgaP/2 = 0,$$
$$ma^2(3Pn^2 + Qn^2) - mgaQ = 0,$$

or
$$\begin{vmatrix} 18an^2 - 15g/2 & 3an^2 \\ 3an^2 & an^2 - g \end{vmatrix} = 0,$$

where P and Q are eliminated. We obtain

$$6a^2n^4 - 17gan^2 + 5g^2 = 0,$$

with roots $n^2 = 5g/2a$, $g/3a$.

In each case, $P/Q = (g - an^2)/3an^2$, with values $-\tfrac{1}{5}$ and $\tfrac{2}{3}$ respectively. These give the ratios of θ to ϕ in each normal mode. $\square$

59

EXERCISES

1. Solve the equations
$$Du+v-w = 0, \quad Dv-w = 0, \quad u+Dw-w = 0,$$
such that $u = 0, v = -1, w = 1$ at $x = 0$.

2. Solve the equations
$$D^2u-6Dv+4u = 9, \quad D^2v+4Du+9v = 0,$$
such that $u = Du = v = 0, Dv = 12$ at $x = 0$.

3. Solve the equations
$$Du-2Dv-2v = 3\sin x+2\cos x, \quad Du+Dv-u+v = 6\sin x+4\cos x,$$
such that $u = v = 0$ at $x = 0$.

4. Solve the equations
$$2D^2u+u-v = 8\cos 2x, \quad 3D^2v-u+2v = -11\cos 2x.$$

5. Two uniform rods AB and BC are smoothly jointed at B and hang freely from a pivot A. AB is of length $5a$ and mass $3m$; BC is of length $3a$ and mass m. The system performs small oscillations under gravity in a vertical plane. Show that the periods of the normal modes are $2\pi\sqrt{(a/g)}$ and $2\pi\sqrt{(5a/g)}$, and find the normal coordinates.

Answers
1. $u = 2\sin x, v = \sin x-\cos x, w = \sin x+\cos x$.
2. $u = 9(1-\cos 6x)/4, \quad v = 2\sin 6x$.
3. $u = 5(e^{2x/3}-\cos x), v = e^{2x/3}-\cos x$.
4. $u = A\cos x+B\sin x+C\cos x/\sqrt{6}+D\sin x/\sqrt{6}-\cos 2x,$
   $v = -A\cos x-B\sin x+\frac{2}{3}C\cos x/\sqrt{6}+\frac{2}{3}D\sin x/\sqrt{6}-\cos 2x.$
5. $\sin\sqrt{(g/a)}t, -5\sin\sqrt{(g/a)}t; 3\sin\sqrt{(g/5a)}t, 5\sin\sqrt{(g/5a)}t$.

Chapter 5

Series Solutions

5.1 Expansion of Certain Functions Some functions may be expanded by first forming a differential equation satisfied by the function, and then by using Leibnitz's theorem to obtain the value of all derivatives at $x = 0$, thus enabling Maclaurin's power series to be written down.

Problem 5.1 Find the power series for the function
$$y = \cos(2n \cos^{-1} x),$$
where n is a positive integer.

Solution. The first derivative is given by
$$y' = -\sin(2n\cos^{-1}x)(-2n)/(1-x^2)^{\frac{1}{2}};$$
squaring
$$(1-x^2)(y')^2 = 4n^2\sin^2(2n\cos^{-1}x)$$
$$= 4n^2(1-y^2).$$

Differentiation yields
$$(1-x^2)2y'y'' - 2x(y')^2 = -4n^2 . 2yy',$$
$$(1-x^2)y'' - xy' + 4n^2 y = 0.$$

The given function is the solution of this equation with
$$y(0) = \cos(2n\cos^{-1}0) = \cos n\pi = (-1)^n,$$
$$y'(0) = 2n\sin(2n\cos^{-1}0) = 0.$$

The differential equation may be differentiated $r-2$ times by Leibnitz's theorem, $y^{(r)}$ denoting the rth derivative, giving
$$(1-x^2)y^{(r)} + (r-2)(-2x)y^{(r-1)} + \tfrac{1}{2}(r-2)(r-3)(-2)y^{(r-2)}$$
$$- xy^{(r-1)} - (r-2)y^{(r-2)} + 4n^2 y^{(r-2)} = 0.$$
When $x = 0$, this simplifies to
$$y^{(r)}(0) = -[4n^2 - (r-2)^2]y^{(r-2)}(0).$$

Since $y'(0) = 0$, it follows that $y^{(3)}(0) = 0$, and also that all the odd derivatives vanish at $x = 0$.

When r is even, $2p$ say, we have
$$y^{2p}(0) = -4[n^2 - (p-1)^2]y^{2(p-1)}(0)$$
$$= (-)^2 4[n^2 - (p-1)^2] . 4[n^2 - (p-2)^2]y^{2(p-2)}(0)$$
$$= (-)^p 4[n^2 - (p-1)^2] . 4[n^2 - (p-2)^2] \ldots 4(n^2)y(0)$$

$$= (-)^p 4^p [n-(p-1)][n-(p-2)]\ldots n$$
$$\times [n+(p-1)][n+(p-2)]\ldots n(-1)^n$$
$$= \begin{cases} 0 \text{ when } p \geqslant n+1, \text{ since a factor } n-n \text{ arises,} \\ (-)^{p+n}4^p \dfrac{n!(n+p-1)!}{(n-p)!(n-1)!} \text{ when } p \leqslant n. \end{cases}$$

Hence the Maclaurin power series is

$$y = \sum_{p=0}^{\infty} \frac{y^{(2p)}(0)}{(2p)!} x^{2p} = \sum_{p=0}^{n} (-)^{p+n}4^p \frac{n(n+p-1)!}{(n-p)!(2p)!} x^{2p}. \qquad \square$$

This is a polynomial of degree $2n$, so the question of convergence does not arise.

5.2 Power-series Solutions

When the coefficients $f(x)$, $g(x)$, $h(x)$ in the equation

$$f(x)D^2y + g(x)Dy + h(x)y = 0$$

are simple polynomials in x, the formal substitution of the assumed power-series solution

$$y = x^c(a_0 + a_1 x + a_2 x^2 + \ldots + a_r x^r + \ldots) \qquad (1)$$

yields, when total coefficients of like powers of x are equated to zero, sufficient equations to determine $c, a_1, a_2, \ldots$. The equation obtained from the lowest index of x determines two values of c, and is known as the *indicial equation*. The resulting series should be tested for convergence by the ratio test. When the two values of c are identical, and sometimes when they differ by an integer, the process fails to produce two independent solutions. A further solution will then involve a logarithm. In some cases, it may be necessary to remove a functional factor from the solution, before evaluating the power series for the remainder.

Problem 5.2 Find two independent power-series solutions of the equation

$$2x(1+x)D^2y + (1+6x)Dy + 2y = 0,$$

expressing them in closed form where possible.

Solution. Substituting the assumed power series (1), and for convenience keeping distinct powers of x on separate lines (each equated to zero), we have (choosing $a_0 = 1$),

$$x^{c-1}: 2c(c-1) \qquad\qquad\qquad\qquad +c \qquad\qquad\qquad = 0,$$
$$x^c \;\; : 2a_1(c+1)c \quad +2c(c-1) \qquad +a_1(c+1)+6c \quad +2 = 0,$$
$$x^{c+1}: 2a_2(c+2)(c+1)+2a_1(c+1)c \quad +a_2(c+2)+6a_1(c+1)+2a_1 = 0,$$
$$x^{c+2}: 2a_3(c+3)(c+2)+2a_2(c+2)(c+1)+a_3(c+3)+6a_2(c+2)+2a_2 = 0,$$

whence
$$c(2c - 1) = 0,$$
$$a_1(c+1)(2c+1) = -2(c+1)^2,$$
$$a_2(c+2)(2c+3) = -2a_1(c+2)^2,$$
$$a_3(c+3)(2c+5) = -2a_2(c+3)^2,$$
$$\cdots \qquad\qquad \cdots$$
$$a_r(c+r)(2c+2r-1) = -2a_{r-1}(c+r)^2.$$

Hence $c = 0$ and $\frac{1}{2}$. When $c = 0$, we have

$$a_1 = -2 \qquad\qquad 5a_3 = -2.3a_2,$$
$$3a_2 = -2.2a_1, \qquad\qquad (2r-1)a_r = -2.ra_{r-1},$$

yielding

$$a_r = -\frac{2ra_{r-1}}{2r-1} = \frac{(-)^2 2^2 r(r-1)a_{r-2}}{(2r-1)(2r-3)} = \frac{(-)^2 2^r . r(r-1)\ldots 1 a_0}{(2r-1)(2r-3)\ldots 1},$$
$$y_1 = 1 - 2x + \frac{1.2}{1.3}(2x)^2 - \frac{1.2.3}{1.3.5}(2x)^3 + \frac{1.2.3.4}{1.3.5.7}(2x)^4 \ldots$$

convergent when $|x| < 1$.

When $c = \frac{1}{2}$, we have

$$a_1 = -3/2, \qquad\qquad 3a_3 = -7a_2/2,$$
$$2a_2 = -5a_1/2, \qquad\qquad ra_r = -\frac{1}{2}(2r+1)a_{r-1},$$

yielding

$$a_r = -(\tfrac{1}{2})\frac{(2r+1)a_{r-1}}{r} = (-)^2(\tfrac{1}{2})^2\frac{(2r+1)(2r-1)a_{r-2}}{r(r-1)}$$
$$= (-)^r(\tfrac{1}{2})^r\frac{(2r+1)(2r-1)\ldots 3a_0}{r(r-1)\ldots 1},$$
$$y_2 = x^{\frac{1}{2}}\left[1 - \frac{3}{2}x + \frac{3.5}{1.2}(\tfrac{1}{2}x)^2 - \frac{3.5.7}{1.2.3}(\tfrac{1}{2}x)^3 \ldots\right] = x^{\frac{1}{2}}(1+x)^{-\frac{3}{2}};$$

again the series is convergent when $|x| < 1$. $\qquad\qquad\qquad\qquad \square$

Problem 5.3 Find the power-series solutions of the equation
$$D^2 y + x^2 y = 0.$$

Solution. Substituting the assumed series (1), we obtain

x^{c-2}: $c(c-1)$ $= 0,$ $\qquad\qquad$ x^{c+1}: $a_3(c+3)(c+2)$ $= 0,$

x^{c-1}: $a_1(c+1)c$ $= 0,$ $\qquad\qquad$ x^{c+2}: $a_4(c+4)(c+3)+1$ $= 0,$

x^c : $a_2(c+2)(c+1) = 0,$ $\qquad\qquad$ x^{c+3}: $a_5(c+5)(c+4)+a_1 = 0.$

The solutions of the indicial equation are $c = 0, 1$, numbers differing by an integer, but in this case two power series emerge. We choose $a_1 = a_2 = a_3 = 0$, the first one not being necessary when $c = 0$, but convenient. This implies that

$$a_5 = a_9 = a_{13} = \ldots = a_6 = a_{10} = a_{14} = \ldots$$
$$= a_7 = a_{11} = a_{15} = \ldots = 0.$$

The remaining equations give

$$a_4 = -\frac{1}{(c+4)(c+3)},$$

$$a_8 = -\frac{a_4}{(c+8)(c+7)} = (-)^2 \frac{1}{(c+8)(c+7)(c+4)(c+3)},$$

and so on. When $c = 0$,

$$a_4 = -\frac{1}{4 \cdot 3}, \quad a_8 = +\frac{1}{8 \cdot 7 \cdot 4 \cdot 3}, \quad a_{12} = -\frac{1}{12 \cdot 11 \cdot 8 \cdot 7 \cdot 4 \cdot 3}.$$

Generally,

$$a_{4r} = (-)^r \, 1/4r(4r-1)(4r-4)(4r-5)\ldots 4 \cdot 3$$
$$= (-)^r \, (4r-2)(4r-3)(4r-6)(4r-7)\ldots 6 \cdot 5 \cdot 2 \cdot 1/(4r)!$$
$$= (-)^r 4^{2r}(r-\tfrac{1}{2})(r-\tfrac{3}{2})\ldots \tfrac{3}{2} \cdot \tfrac{1}{2}(r-\tfrac{3}{4})(r-\tfrac{7}{4})\ldots \tfrac{5}{2} \cdot \tfrac{1}{4}/(4r)!$$
$$= (-)^r 4^{2r} \, (r-\tfrac{1}{2})! \, (r-\tfrac{3}{4})!/(-\tfrac{1}{2})! \, (-\tfrac{3}{4})! \, (4r)!$$

When $c = 1$,

$$a_4 = -\frac{1}{5 \cdot 4}, \quad a_8 = +\frac{1}{9 \cdot 8 \cdot 5 \cdot 4}, \quad a_{12} = -\frac{1}{13 \cdot 12 \cdot 9 \cdot 8 \cdot 5 \cdot 4}.$$

Generally, the coefficient a_{4r} may be rearranged to have the form

$$a_{4r} = (-)^r 4^{2r} \frac{(r-\tfrac{1}{4})! \, (r-\tfrac{1}{2})!}{(-\tfrac{1}{4})! \, (-\tfrac{1}{2})! \, (4r+1)!}. \qquad \square$$

Both series are convergent for all values of x.

Problem 5.4 Examine the differential equation

$$D^2 y + (x^2 - a^2) y = 0.$$

Solution. To simplify this equation, introduce the change of dependent variable

$$y = e^{-\frac{1}{2}ix^2} u.$$

(In elementary work, the student could not guess this substitution; it is really connected with the asymptotic forms of the solution for large $|x|$.) Upon differentiation twice, substitution, and cancellation of the exponential

64

factor, we obtain the equation for u:
$$D^2u - 2ix\,Du - iu - a^2u = 0.$$
To find the power-series solutions, we assume formally
$$u = x^c(1 + a_1 x^2 + a_2 x^4 + \ldots),$$
since inspection shows that all indices must differ from each other by an even integer. Substitution yields

$$x^{c-2}: \qquad\qquad c(c-1) = 0,$$
$$x^c: \qquad (c+2)(c+1)a_1 = 2ic + i + a^2,$$
$$x^{c+2}: \qquad (c+4)(c+3)a_2 = [2i(c+2) + i + a^2]a_1,$$
$$x^{c+4}: \qquad (c+6)(c+5)a_3 = [2i(c+4) + i + a^2]a_2.$$

When $c = 0$, we have
$$a_1 = (i + a^2)/2!, \qquad a_2 = (5i + a^2)(i + a^2)/4!,$$
$$a_3 = (9i + a^2)(5i + a^2)(i + a^2)/6!;$$
and when $c = 1$, we have
$$a_1 = (3i + a^2)/3!, \qquad a_2 = (7i + a^2)(3i + a^2)/5!,$$
$$a_3 = (11i + a^2)(7i + a^2)(3i + a^2)/7!.$$

The general power-series solution is therefore
$$y = Ae^{-\frac{1}{2}ix^2}\left[1 + \frac{i + a^2}{2!}x^2 + \frac{(5i + a^2)(i + a^2)}{4!}x^4 + \ldots\right]$$
$$+ Be^{-\frac{1}{2}ix^2}x\left[1 + \frac{3i + a^2}{3!}x^2 + \frac{(7i + a^2)(3i + a^2)}{5!}x^4 + \ldots\right]$$
$$\equiv Ae^{-\frac{1}{2}ix^2}p_1 + Be^{-\frac{1}{2}ix^2}xp_2, \quad \text{say.}$$

Similarly, if we place $y = e^{\frac{1}{2}ix^2}v$, we obtain
$$D^2v + 2ix\,Dv + iv - a^2v = 0,$$
possessing the general solution
$$y = Ce^{\frac{1}{2}ix^2}\left[1 - \frac{i - a^2}{2!}x^2 + \frac{(5i - a^2)(i - a^2)}{4!}x^4 - \ldots\right]$$
$$+ De^{\frac{1}{2}ix^2}x\left[1 - \frac{3i - a^2}{3!}x^2 + \frac{(7i - a^2)(3i - a^2)}{5!}x^4 - \ldots\right]$$
$$\equiv Ce^{\frac{1}{2}ix^2}q_1 + De^{\frac{1}{2}ix^2}xq_2, \quad \text{say.}$$

All series are convergent for all values of x. If, in these two distinct forms of solution, we put $x = 0$, we find that $A = C$, while if we first differentiate with respect to x, and put $x = 0$, we find $B = D$, yielding the identities
$$e^{-\frac{1}{2}ix^2}p_1 \equiv e^{\frac{1}{2}ix^2}q_1, \qquad e^{-\frac{1}{2}ix^2}p_2 \equiv e^{\frac{1}{2}ix^2}q_2. \qquad \square$$

65

5.3 Solutions Involving Logarithms In other cases, we find the values of the coefficients $a_1, a_2, a_3, \ldots$ in terms of c, without inserting the actual values of c, namely c_1 and c_2, found from the indicial equation. Let $p(x, c)$ denote the power series formed by these coefficients.

When $c_1 = c_2$, two independent solutions of the differential equation are

$$y_1 = p(x, c)\Big|_{c=c_1}, \qquad y_2 = \frac{\partial p(x, c)}{\partial c}\Big|_{c=c_1}.$$

When c_1 and c_2 differ by an integer, and in cases when only one power series is produced, it is the larger root, c_1 say, that yields a power-series solution as before. Previously, we have taken a_0 to equal unity, but now take a_0 to equal $c - c_2$. Evaluate all the coefficients $a_1, a_2, a_3, \ldots$ in terms of c and a_0 (a factor $c - c_2$ will cancel in each term), giving a power series $p(x, c)$. Then two independent solutions of the equation are

$$y_1 = p(x, c)\Big|_{c=c_1}, \qquad y_2 = \frac{\partial p(x, c)}{\partial c}\Big|_{c=c_2}.$$

It will be found that $p(x, c_2)$ yields the same series as $p(x, c_1)$. It should be pointed out that differentiation with respect to c may introduce very awkward terms.

Problem 5.5 Find solutions of the equation

$$x^2 D^2 y + (x^2 - 5x) Dy + (9 - 3x) y = 0.$$

Solution. Substituting the assumed power series (1), we obtain

x^c:	$c(c-1)$	$-5c$	$+9$	$= 0,$
x^{c+1}:	$(c+1)ca_1 + c$	$-5(c+1)a_1 + 9a_1 - 3$		$= 0,$
x^{c+2}:	$(c+2)(c+1)a_2 + (c+1)a_1 - 5(c+2)a_2 + 9a_2 - 3a_1$			$= 0,$
x^{c+3}:	$(c+3)(c+2)a_3 + (c+2)a_2 - 5(c+3)a_3 + 9a_3 - 3a_2$			$= 0,$

whence $(c-3)^2 = 0,$ $(c-2)^2 a_1 + (c-3) = 0,$

$(c-1)^2 a_2 + (c-2) a_1 = 0,$ $c^2 a_3 + (c-1) a_2 = 0.$

The indicial equation has two equal roots $c = 3$. In terms of c, we have

$$a_1 = -\frac{c-3}{(c-2)^2}, \qquad a_2 = \frac{c-3}{(c-1)^2(c-2)}, \qquad a_3 = -\frac{c-3}{c^2(c-1)(c-2)},$$

$$a_r = (-)^r \frac{c-3}{(c-3+r)^2(c-4+r)\ldots(c-1)(c-2)}.$$

When $c = 3$, the only non-vanishing coefficient is the first, yielding the solution $y_1 = x^3$.

Secondly, we form the series

$$p(x, c) = x^c \left[1 + \sum_{r=1}^{\infty} (-)^r \frac{(c-3)x^r}{(c-3+r)^2 (c-4+r)\ldots(c-1)(c-2)} \right].$$

The second solution is then $y_2 = [\partial p(x, c)/\partial c]_{c=3}$. Since $x^c \equiv e^{c \log x}$, we have

$$\frac{\partial p}{\partial c} = \log x \, e^{c \log x} \times \text{series} + x^c \frac{\partial}{\partial c} \text{ series,}$$

becoming, when $c = 3$,

$$y_2 = x^3 \log x + x^3 \left[\frac{\partial}{\partial c} \sum_{r=1}^{\infty} (-)^r \frac{(c-3)x^r}{(c-3+r)^2(c-4+r)\ldots(c-2)} \right]_{c=3}.$$

Clearly only the numerator need be differentiated to yield non-zero terms; the square bracket becomes

$$\left[\sum_{r=1}^{\infty} (-)^r \frac{x^r}{(c-3+r)^2 (c-4+r)\ldots(c-2)} \right]_{c=3} = \sum_{r=1}^{\infty} \frac{(-)^r x^r}{r\,r!}.$$

Hence $\qquad\qquad y_2 = x^3 \log x + x^3 \sum_{r=1}^{\infty} (-)^r x^r/r\,r!.$ $\qquad\qquad$ ☐

Problem 5.6 Find two expansions for solutions of Bessel's equation of order 1.

Solution. We are given the equation

$$D^2 y + \frac{Dy}{x} + \left(1 - \frac{1}{x^2} \right) y = 0,$$

and we substitute the power series (1) into the corresponding equation

$$x^2 D^2 y + x\, Dy + (x^2 - 1) y = 0,$$

thereby obtaining the equations

$$\begin{aligned}
x^c &: c(c-1) &+ c &\quad -1 = 0, \\
x^{c+1} &: (c+1)ca_1 &+ (c+1)a_1 &\quad -a_1 = 0, \\
x^{c+2} &: (c+2)(c+1)a_2 + (c+2)a_2 + 1 &\quad -a_2 = 0, \\
x^{c+3} &: (c+3)(c+2)a_3 + (c+3)a_3 + a_1 &\quad -a_3 = 0.
\end{aligned}$$

Hence,

$$c^2 - 1 = 0, \qquad\qquad [(c+2)^2 - 1] a_2 = -1,$$
$$[(c+1)^2 - 1] a_1 = 0, \qquad\qquad [(c+3)^2 - 1] a_3 = -a_1,$$

and so on.

67

The roots of the indicial equation are $c = \pm 1$, differing by an integer. This demands that $a_1 = 0$, and hence all the odd coefficients vanish. Then

$$a_2 = -1/(c+1)(c+3),$$
$$a_4 = +1/(c+3)(c+1)(c+5)(c+3),$$
$$a_{2r} = (-1)^r/(c+2r-1)\ldots(c+1)(c+2r+1)\ldots(c+3).$$

When $c = 1$,

$$a_{2r} = \frac{(-1)^r}{(2r)(2r-2)\ldots2(2r+2)(2r)\ldots4} = \frac{(-1)^r}{2^{2r}r!(r+1)!},$$

yielding a series solution

$$y_1 = x \sum_{r=0}^{\infty} (-\tfrac{1}{4}x^2)^r/r!\,(r+1)!,$$

convergent for all values of x.

We cannot insert $c = -1$, since all coefficients contain the factor $(c+1)$ in the denominator. We therefore form the series $p(x,c)$ with the above coefficients in terms of c, but having multiplied through by the offending factor $(c+1)$. We have

$$p(x,c) = x^c\left[(c+1) - \frac{x^2}{c+3} + \ldots\right.$$
$$\left. + \frac{(-x^2)^r}{(c+2r-1)\ldots(c+3)(c+2r+1)\ldots(c+3)} + \ldots\right].$$

Inserting $c = -1$, we obtain a series proportional to y_1, namely $-\tfrac{1}{2}y_1$ by inspection.

The second solution is given by $y_2 = [\partial p(x,c)/\partial c]_{c=-1}$. Differentiation of the term x^c yields

$$x^{-1}\log x\left[0 - \frac{x^2}{2} + \frac{x^4}{2.4.2} - \ldots\right] \equiv -\tfrac{1}{2}y_1 \log x.$$

Differentiation of the square bracket yields the initial terms

$$x^{-1}\left[1 + \frac{x^2}{(c+3)^2} + \ldots\right]_{c=-1} \equiv x^{-1}(1 + \tfrac{1}{4}x^2 + \ldots).$$

We designate the denominator of the general term by $d(c)$, where
$$d(-1) = (2r-2)\ldots2(2r)\ldots2 = 2^{2r-1}(r-1)!\,r!.$$

68

Differentiation of the general term in the square bracket then gives:

$$(-x^2)^r \left[\frac{\partial}{\partial c} \frac{1}{d(c)}\right]_{c=-1}$$

$$= -(-x^2)^r \left[\left(\frac{1}{c+2r-1} + \ldots + \frac{1}{c+3} + \frac{1}{c+2r+1} + \ldots + \frac{1}{c+3}\right)\frac{1}{d(c)}\right]_{c=-1}$$

$$= -2\left[-\frac{x^2}{4}\right]^r \left[\frac{1}{2r-2} + \frac{1}{2r-4} + \ldots + \frac{1}{2} + \frac{1}{2r} + \frac{1}{2r-2} + \ldots + \frac{1}{2}\right]\frac{1}{r!(r-1)!}$$

$$= -\left[-\frac{x^2}{4}\right]^r \left[\frac{1}{r} + \frac{2}{r-1} + \frac{2}{r-2} + \ldots + \frac{2}{3} + \frac{2}{2} + 2\right]\frac{1}{r!(r-1)!}.$$

Hence the general form for y_2 is

$$y_2 = -\frac{y_1}{2}\log x + \frac{1}{x}\left[1 + \frac{x^2}{4} - \sum_{r=2}^{\infty} \left(\frac{1}{r} + \frac{2}{r-1} + \ldots + \frac{2}{2} + 2\right)\frac{(-\frac{1}{4}x^2)^r}{r!(r-1)!}\right].$$

□

EXERCISES

1. Find two independent power-series solutions of the equation
$$2x\,D^2 y + (3-2x)\,Dy - 2y = 0.$$

2. Find two independent power-series solutions of the Airy equation
$$D^2 y = xy.$$

3. Solve the following equation (a logarithm will appear in the solution)
$$D^2 y + x^{-1}\,Dy + xy = 0.$$

Answers

1. $y_1 = e^x/\sqrt{x}$, $y_2 = 1 + \dfrac{2x}{3} + \dfrac{(2x)^2}{3.5} + \dfrac{(2x)^3}{3.5.7} + \dfrac{(2x)^4}{3.5.7.9} + \ldots$

2. $y_1 = 1 + \dfrac{x^3}{2.3} + \dfrac{x^6}{2.3.5.6} + \dfrac{x^9}{2.3.5.6.8.9} + \ldots,$

   $y_2 = x + \dfrac{x^3}{3.4} + \dfrac{x^7}{3.4.6.7} + \dfrac{x^{10}}{3.4.6.7.9.10} + \ldots.$

3. $y_1 = 1 - \dfrac{x^3}{3^2} + \dfrac{x^6}{3^4(2!)^2} - \dfrac{x^9}{3^6(3!)^2} + \ldots,$

   $y_2 = y_1 \log x + 2\left[\dfrac{x^3}{3^3} - \dfrac{(1+\frac{1}{2})x^6}{3^5(2!)^2} + \dfrac{(1+\frac{1}{2}+\frac{1}{3})x^9}{3^7(3!)^2} - \ldots\right].$

Chapter 6

The Laplace Transform

6.1 Laplace Transforms and their Inverses The Laplace transform, $\mathcal{L}(y) \equiv \bar{y}(p)$, of a function $y(x)$ is obtained by multiplying by e^{-px}, and integrating with respect to x between the limits 0 and ∞, provided that a range of values of p exists for which the integral converges. Thus

$$\mathcal{L}(y) \equiv \bar{y}(p) \equiv \int_0^\infty e^{-px} y(x)\, dx.$$

Tables giving the unique correspondence $y(x) \leftrightarrow \bar{y}(p)$ may be used in elementary work; a small selection is given below.

$y(x)$	$\bar{y}(p)$	$y(x)$	$\bar{y}(p)$
$x^n, n =$ positive integer	$n!/p^{n+1}$	$\cos ax$	$p/(p^2 - a^2)$
$x^n e^{-ax}$	$n!/(p+a)^{n+1}$	$\sinh ax$	$a/(p^2 - a^2)$
$\cos ax$	$p/(p^2 + a^2)$	$x \cos ax$	$(p^2 - a^2)/(p^2 + a^2)^2$
$\sin ax$	$a/(p^2 + a^2)$	$x \sin ax$	$2ap/(p^2 + a^2)^2$

The following should be noted:
$$\mathcal{L}[x^n y(x)] = (-1)^n d^n \bar{y}(p)/dp^n,$$
$$\mathcal{L}[e^{-ax} y(x)] = \bar{y}(p+a),$$
$$\mathcal{L}(Dy) = p\bar{y}(p) - y(0),$$
$$\mathcal{L}(D^2 y) = p^2 \bar{y}(p) - Dy(0) - py(0).$$

Partial fractions are usually necessary in evaluating $y(x)$ from given $\bar{y}(p)$. In more comprehensive work, $\bar{y}(p)$ may be regarded as a function of the complex variable p. Then

$$y(x) = \frac{1}{2\pi i} \int_{\tau - i\infty}^{\tau + i\infty} e^{px} \bar{y}(p)\, dp,$$

where $\tau >$ real part of all singularities of $\bar{y}(p)$.

Problem 6.1 Prove the result that
$$\mathcal{L}[e^{-ax} y(x)] = \bar{y}(p+a).$$

Solution. It is given that
$$\mathcal{L}[y(x)] \equiv \bar{y}(p) = \int_0^\infty e^{-px} y(x)\, dx;$$

hence
$$\mathcal{L}[e^{-ax}y(x)] = \int_0^\infty e^{-px}[e^{-ax}y(x)]\,dx$$
$$= \int_0^\infty e^{-(p+a)x}y(x)\,dx = \bar{y}(p+a). \qquad \square$$

Problem 6.2   Prove the result that
$$\mathcal{L}[x^n y(x)] = (-1)^n d^n\,\bar{y}(p)/dp^n.$$

Solution. The defining integral for $\bar{y}(p)$ is differentiated n times with respect to p, differentiation under the sign of integration being assumed permissible. We obtain

$$d^n\,\bar{y}(p)/dp^n = \int_0^\infty (d^n/dp^n)\,e^{-px}y(x)\,dx$$
$$= \int_0^\infty (-x)^n e^{-px}y(x)\,dx;$$

hence
$$(-1)^n d^n\bar{y}(p)/dp^n = \int_0^\infty e^{-px}[x^n y(x)]\,dx,$$

and the result follows. $\qquad \square$

Since $\mathcal{L}(1) = 1/p$ and $\mathcal{L}(e^{-ax}) = 1/(p+a)$, the first two entries in the table follow by this result.

Problem 6.3   Prove the result that
$$\mathcal{L}(D^2 y) = p^2\,\bar{y}(p) - Dy(0) - py(0).$$

Solution. Firstly, if z is a function of x,

$$\mathcal{L}(Dz) = \int_0^\infty e^{-px}(Dz)\,dx$$
$$= \left[ze^{-px}\right]_0^\infty - \int_0^\infty (-pe^{-px})z\,dx \qquad \text{(by parts)}$$
$$= -z(0) + p\,\mathcal{L}(z),$$

where z must be such that $ze^{-px} \to 0$ as $x \to \infty$. Then if $z = Dy$,

$$\mathcal{L}(D^2 y) = -Dy(0) + p\,\mathcal{L}(Dy)$$
$$= -Dy(0) + p[-y(0) + p\,\mathcal{L}(y)]$$
$$= p^2\bar{y} - Dy(0) - py(0). \qquad \square$$

Problem 6.4   Find the Laplace transform of
$$y(x) = 8e^{-\frac{1}{2}x}[\cos\tfrac{1}{2}\sqrt{(3)}x + \sqrt{(3)}\sin\tfrac{1}{2}\sqrt{(3)}x] - 3e^{-x}(3 + 6x + 2x^2) + e^x.$$

Solution. The Laplace Transform of $\cos\tfrac{1}{2}\sqrt{(3)}x$ is given by $p/(p^2 + \tfrac{3}{4})$, so

71

the transform of $e^{-\frac{1}{2}x}\cos\frac{1}{2}\sqrt{(3)}$ is

$$\frac{(p+\frac{1}{2})}{(p+\frac{1}{2})^2+\frac{3}{4}} \equiv \frac{p+\frac{1}{2}}{p^2+p+1}.$$

Totally, we have

$$\bar{y} = \frac{8(p+\frac{1}{2})}{p^2+p+1}+\frac{8\sqrt{(3)}\cdot\frac{1}{2}\sqrt{(3)}}{(p+\frac{1}{2})^2+\frac{3}{4}}-3\left[\frac{3}{p+1}+\frac{6}{(p+1)^2}+\frac{2.2}{(p+1)^3}\right]+\frac{1}{p-1}$$
$$= 24/(p+1)^3(p^3-1).$$ □

Problem 6.5 Find the inverse Laplace transform of
$$\bar{y}(p) = 2(p^3+4p^2+6p+2)/[(p+1)^4-1].$$

Solution. Since the denominator factorizes thus:
$$(p+1)^4-1 \equiv [(p+1)^2-1][(p+1)^2+1] \equiv p(p+2)(p^2+2p+2),$$
we assume the partial fractions
$$\bar{y}(p) \equiv \frac{A}{p}+\frac{B}{p+2}+\frac{Cp+D}{p^2+2p+2}.$$

After cross-multiplication,
$$2(p^3+4p^2+6p+2)$$
$$\equiv A(p+2)(p^2+2p+2)+Bp(p^2+2p+2)+(Cp+D)p(p+2).$$

When $p = 0$, we see that $A = 1$. When $p = -2$, we see immediately that $B = 1$. Identifying coefficients of p^3, we obtain $2 = A+B+C$, showing that $C = 0$. Finally, the coefficients of p^2 yield $8 = 4A+2B+D$, giving $D = 2$. Hence

$$\bar{y}(p) = \frac{1}{p}+\frac{1}{p+2}+\frac{2}{p^2+2p+2}$$
$$= \frac{1}{p}+\frac{1}{p+2}+\frac{2}{(p+1)^2+1},$$

where any quadratic denominators must be rearranged to equal the square of a linear expression in p plus a constant. Hence

$$y(x) = 1+e^{-2x}+2e^{-x}\sin x,$$

where the tables have been used backwards for the respective entries. □

Problem 6.6 Find the inverse Laplace transform of
$$\bar{y}(p) = (21p^2-2)/(4p^2+4p+10)(3p-8).$$

Solution. In terms of partial fractions, let
$$\bar{y}(p) \equiv \frac{Ap+B}{4p^2+4p+10}+\frac{C}{3p-8},$$

or
$$21p^2 - 2 \equiv (Ap + B)(3p - 8) + C(4p^2 + 4p + 10).$$

Comparing like powers of p, we obtain the equations
$$21 = 3A + 4C, \quad 0 = -8A + 3B + 4C, \quad -2 = -8B + 10C,$$

with the solution $A = 3$, $B = 4$, $C = 3$.

The quadratic term in the denominator must be rearranged so as to contain a square of a linear expression in p; the coefficients of p in the denominators must be rearranged to equal unity, and the numerator of a quadratic denominator must be arranged to contain the same linear term as in the denominator. Thus

$$\bar{y}(p) = \frac{3p + 4}{(2p + 1)^2 + 9} + \frac{3}{3p - 8}$$

$$\equiv \frac{\frac{3}{4}p + 1}{(p + \frac{1}{2})^2 + 9/4} + \frac{1}{p - 8/3} \equiv \frac{\frac{3}{4}(p + \frac{1}{2}) + \frac{5}{8}}{(p + \frac{1}{2})^2 + 9/4} + \frac{1}{p - 8/3}.$$

The use of the table then yields

$$y(x) = e^{-\frac{1}{2}x}[\tfrac{3}{4}\cos(3x/2) + (5/12)\sin(3x/2)] + e^{8x/3}.$$

When evaluating inverse transforms by complex contour integration, the straight line path $\tau - i\infty$ to $\tau + i\infty$ is closed in the complex p-plane by the infinite part-circle lying on its left. The integral of $e^{px}\bar{y}(p)$ taken around the part-circle vanishes when

$$|\bar{y}(p)| < AR^{-a}, \quad \text{where} \quad p = Re^{i\theta}, \ -\pi \leqslant \theta \leqslant \pi, R > R_0, a > 0,$$

$x > 0$. This condition is always satisfied when $\bar{y}(p)$ is a ratio of polynomials in p, the degree of the numerator being less than the degree of the denominator. Then

$$y(x) = \frac{1}{2\pi i} \int_{\tau - i\infty}^{\tau + i\infty} e^{px}\, \bar{y}(p)\, dp = \sum \text{residues of } e^{px}\bar{y}(p).$$

The residue of $e^{px}\bar{y}(p)$ at the singularity $p = \alpha$ is

$$\lim_{p \to \alpha}[(p - \alpha)e^{px}\,\bar{y}(p)] \equiv \left[\left(\frac{d}{dp}\frac{1}{\bar{y}(p)}\right)^{-1}\right]_{p = \alpha}$$

when the singularity is of order one. When the pole is of order n, the coefficient of $1/(p - \alpha)$ occurring in the expansion of $e^{px}\bar{y}(p)$ in terms of $(p - \alpha)$ must be calculated. Its value is

$$\frac{1}{(n - 1)!}\frac{d^{n-1}}{dp^{n-1}}[(p - \alpha)^n e^{px}\,\bar{y}(p)]_{p = \alpha}.$$

Problem 6.7 Find the inverse Laplace transform of
$$\bar{y}(p) = (p^2 - 3p + 3)/(p - 1)(p - 2)^2$$
by the method of complex contour integration.

Solution. The poles of $\bar{y}(p)$ are given by $p = 1, 2$, so taking $\tau = 3$, we evaluate

$$y(x) = \frac{1}{2\pi i} \int_{3-i\infty}^{3+i\infty} \frac{e^{px}(p^2 - 3p + 3)}{(p-1)(p-2)^2} \, dp.$$

Completing the closed contour as described, we see that

$$y(x) = \text{residue at } 1 + \text{residue at } 2.$$

The residue at $p = 1$ (a pole of order 1) equals

$$\left[\frac{e^{px}(p^2 - 3p + 3)}{(p-2)^2} \right]_{p=1} = e^x.$$

The residue at $p = 2$ (a second order pole) equals

$$\frac{d}{dp}\left[(p-2)^2 e^{px} \bar{y}(p)\right]_{p=2} = \frac{d}{dp}\left[\frac{e^{px}(p^2 - 3p + 3)}{p-1} \right]_{p=2}$$

$$= \left[\frac{pe^{px}(p^2 - 3p + 3)}{p-1} + \frac{e^{px}(2p - 3)}{p-1} \right.$$

$$\left. - \frac{e^{px}(p^2 - 3p + 3)}{(p-1)^2} \right]_{p=2}$$

$$= e^{2x}. \qquad \square$$

Hence $y(x) = e^x + e^{2x}$.

Problem 6.8   Find the inverse Laplace transform of

$$\bar{y}(p) = (6p^2 - 2p^4)/(p^2 + 1)^3$$

by the method of complex contour integration.

Solution. Following general theory, $\bar{y}(p)$ having poles $p = \pm i$ each of order three, we see that

$$y(x) = \sum \text{residues of } \frac{e^{px}(6p^2 - 2p^4)}{(p^2 + 1)^3} \text{ at } p = \pm i$$

$$= (\text{residue at } p = +i) + \text{its complex conjugate (c.c.)}$$

$$= \frac{1}{2}\left[\frac{d^2}{dp^2}(p - i)^3 \frac{e^{px}(6p^2 - 2p^4)}{(p^2 + 1)^3} \right]_{p=i} + \text{c.c.}$$

$$= D^2[e^{px}(3p^2 - p^4)/(p+i)^3]_{p=i} + \text{cc} \qquad (D \equiv d/dp)$$

$$= e^{px}(D+x)^2 \left[\frac{3p^2 - p^4}{(p+i)^3} \right]_{p=i} + \text{c.c.} \qquad \begin{array}{l}\text{(by the rule for extracting a} \\ \text{factor } e^{px} \text{ from the operator} \\ D^2)\end{array}$$

$$= \left[e^{px} \left\{ \frac{6-12p^2}{(p+i)^3} - \frac{6(6p-4p^3)}{(p+i)^4} + \frac{12(3p^2-p^4)}{(p+i)^5} \right. \right.$$
$$\left. \left. + 2x \left(\frac{6p-4p^3}{(p+i)^3} - \frac{3(3p^2-p^4)}{(p+i)^4} \right) + x^2 \left(\frac{3p^2-p^4}{(p+i)^3} \right) \right\} \right]_{p=i} + \text{c.c.}$$
$$= e^{ix}(-x - \tfrac{1}{2}ix^2) + \text{c.c.} = x^2 \sin x - 2x \cos x. \qquad \square$$

6.2 Solution of Differential Equations

The Laplace transform of the differential equation is written down, using the given initial conditions; $\bar{y}(p)$ is found from which $y(x)$ is deduced using either the tables or complex contour integration.

Problem 6.9 Solve the equation $D^2y - Dy - 2y = 0$, with $y(0) = 1$, $Dy(0) = 5$.

Solution. The Laplace transform is
$$(p^2\bar{y} - 5 - p) - (p\bar{y} - 1) - 2\bar{y} = 0,$$
$$\bar{y} = \frac{p+4}{p^2-p-2} = \frac{2}{p-2} - \frac{1}{p+1}.$$
Hence $\qquad y(x) = 2e^{-2x} - e^{-x}.$ $\qquad \square$

Problem 6.10 Solve the equation $D^3y - D^2y - Dy + 1 = 0$, with $y(0) = 2$, $Dy(0) = 1$, $D^2y(0) = 4$.

Solution. The Laplace transform is
$$(p^3\bar{y} - 4 - p - 2p^2) - (p^2\bar{y} - 1 - 2p) - (p\bar{y} - 2) + \bar{y} = 0,$$
$$\bar{y} = \frac{2p^2-p+1}{p^3-p^2-p+1} = \frac{1}{p+1} + \frac{1}{p-1} + \frac{1}{(p-1)^2}.$$
Hence $\qquad y(x) = e^{-x} + e^x + xe^x.$ $\qquad \square$

Problem 6.11 Solve the equation $D^2y - 3Dy + y = 1 - 4x + 2x^2$, with $y(0) = 4$, $Dy(0) = 5$.

Solution. The Laplace transform is
$$(p^2\bar{y} - 5 - 4p) - 3(p\bar{y} - 4) + 2\bar{y} = \frac{1}{p} - \frac{4}{p^2} + \frac{4}{p^3},$$
$$\bar{y} = \frac{4p-7}{(p-1)(p-2)} + \frac{p^2-4p+4}{p^3(p-1)(p-2)}$$
$$= \frac{2}{p-1} + \frac{1}{p-2} + \frac{1}{p} + \frac{1}{p^2} + \frac{2}{p^3}$$
upon collecting together the partial fractions. Hence
$$y(x) = 2e^x + e^{2x} + 1 + x + x^2.$$ $\qquad \square$

Problem 6.12 Solve the equation $D^2y - 4Dy + 3y = 6e^{4x}$, with $y(0) = 2$, $Dy(0) = 6$.

Solution. The Laplace transform is

$$(p^2\bar{y} - 6 - 2p) - 4(p\bar{y} - 2) + 3\bar{y} = 6/(p-4),$$

$$\bar{y} = \frac{2p-2}{(p-1)(p-3)} + \frac{6}{(p-1)(p-3)(p-4)}$$

$$= \frac{1}{p-1} - \frac{1}{p-3} + \frac{2}{p-4}.$$

Hence $\qquad y(x) = e^x - e^{3x} + 2e^{4x}.$ $\qquad\square$

Problem 6.13 Solve the equation $D^2y - 2Dy + y = 6xe^x$, with $y(0) = 1$, $Dy(0) = 0$.

Solution. The Laplace transform is

$$(p^2\bar{y} - p) - 2(p\bar{y} - 1) + \bar{y} = 6/(p-1)^2,$$

$$\bar{y} = \frac{p-2}{(p-1)^2} + \frac{6}{(p-1)^4} = \frac{1}{p-1} - \frac{1}{(p-1)^2} + \frac{6}{(p-1)^4}.$$

Hence $\qquad y(x) = e^x - xe^x + x^3e^x.$ $\qquad\square$

Problem 6.14 Solve the equation $D^2y + 4y = 10\sin 3x - 5\cos 3x$, with $y(0) = 2$, $Dy(0) = -4$.

Solution. The Laplace transform is

$$(p^2\bar{y} + 4 - 2p) + 4\bar{y} = \frac{30}{p^2+9} - \frac{5p}{p^2+9},$$

$$\bar{y} = \frac{2p-4}{p^2+4} + \frac{30-5p}{(p^2+4)(p^2+9)}$$

$$= \left(\frac{2p}{p^2+4} - \frac{4}{p^2+4}\right) + \left(-\frac{p}{p^2+4} + \frac{6}{p^2+4} + \frac{p}{p^2+9} - \frac{6}{p^2+9}\right)$$

$$= \frac{p}{p^2+4} + \frac{2}{p^2+4} + \frac{p}{p^2+9} - \frac{6}{p^2+9};$$

hence $\qquad y(x) = \cos 2x + \sin 2x + \cos 3x - 2\sin 2x.$ $\qquad\square$

Problem 6.15 Solve the equation $D^2y + 9y = 6\cos 3x$, with $y(0) = 0$, $D(y) = 3$.

Solution. The Laplace transform is

$$(p^2\bar{y} - 3) + 9\bar{y} = 6p/(p^2+9),$$

$$\bar{y} = 3/(p^2+9) + 6p/(p^2+9)^2;$$

hence $\qquad y(x) = \sin 3x + x\sin 3x.$ $\qquad\square$

Problem 6.16 Solve the equation $D^2y - 2Dy + 2y = 2e^x \sin x$, with $y(0) = Dy(0) = 0$.

Solution. The Laplace transform is

$$p^2\bar{y} - 2p\bar{y} + 2\bar{y} = 2/[(p-1)^2 + 1],$$
$$\bar{y} = 2/[(p-1)^2 + 1]^2.$$

If $p - 1 = s$, this must be arranged in terms of the table entries

$$\frac{s}{s^2+1}, \quad \frac{1}{s^2+1}, \quad \frac{s^2-1}{(s^2+1)^2}, \quad \frac{2s}{(s^2+1)^2}.$$

Inspection shows that

$$\bar{y} = \frac{(s^2+1) - (s^2-1)}{(s^2+1)^2}$$

$$= \frac{1}{s^2+1} - \frac{s^2-1}{(s^2+1)^2}$$

$$= \frac{1}{(p-1)^2+1} - \frac{(p-1)^2-1}{[(p-1)^2+1]^2};$$

hence

$$y(x) = e^x(\sin x - x \cos x).$$ □

6.3 Simultaneous Differential Equations Linear simultaneous differential equations in the dependent variables $u, v, w, \ldots$ (x being the independent variable) may be solved by taking the Laplace transform of all the given equations (using the given initial conditions), solving for $\bar{u}, \bar{v}, \bar{w}, \ldots$, and hence finding $u, v, w, \ldots$.

Problem 6.17 Solve the equations

$$2Du + 3Dv + 8u = e^x,$$
$$Du + Dv - v = 0,$$

with $u = 1, v = 0$ when $x = 0$.

Solution. The Laplace transforms of the equations are

$$2(p\bar{u} - 1) + 3(p\bar{v} + 0) - 8\bar{u} = 1/(p-1),$$
$$(p\bar{u} - 1) + (p\bar{v} - 0) - \bar{v} = 0,$$

or

$$(2p+8)\bar{u} + 3p\bar{v} = 2 + 1/(p-1) = (2p-1)/(p-1),$$
$$p\bar{u} + (p-1)\bar{v} = 1.$$

Their solution using determinants is

$$\frac{\bar{u}}{\begin{vmatrix} 3p & -(2p-1)/(p-1) \\ p-1 & -1 \end{vmatrix}} = \frac{-\bar{v}}{\begin{vmatrix} 2p+8 & -(2p-1)/(p-1) \\ p & -1 \end{vmatrix}} = \frac{1}{\begin{vmatrix} 2p+8 & 3p \\ p & p-1 \end{vmatrix}},$$

F

yielding after some manipulation,

$$\bar{u} = \frac{-3/2}{p-2} + \frac{5/2}{p-4}, \qquad \bar{v} = \frac{1/3}{p-1} + \frac{3}{p-2} - \frac{10/3}{p-4};$$

$$u(x) = -3e^{2x}/2 + 5e^{4x}/2, \quad v(x) = e^x/3 + 3e^{2x} - 10e^{-4x}/3. \qquad \square$$

Problem 6.18   Solve the equations

$$u + Dv + D^2w = 9x, \quad D^2u + v + Dw = 4x^2, \quad Du + D^2v + w = 3 + x^3,$$

with $u(0) = v(0) = w(0) = 0$, $Du(0) = 1$, $Dv(0) = Dw(0) = 0$.

Solution. The Laplace transforms are

$$\bar{u} + p\bar{v} + p^2\bar{w} = 9/p^2, \tag{1}$$

$$p^2\bar{u} - 1 + \bar{v} + p\bar{w} = 8/p^3, \tag{2}$$

$$p\bar{u} + p^2\bar{v} + \bar{w} = 3/p + 6/p^4. \tag{3}$$

Take $(1) - p(2)$: $\qquad (1-p^3)\bar{u} = 9/p^2 - p(8/p^3 + 1)$,

$$\bar{u} = 1/p^2,$$

giving $u = x$.

Take $(2) - p(3)$. This gives $\bar{v} = 2/p^3$ and $v = x^2$.

Take $(3) - p(1)$. This gives $\bar{w} = 6/p^4$ and $w = x^3$. $\qquad \square$

6.4 Laplace Transforms of Further Functions   *Impulse function.* The notation $\delta(x-a)$ denotes an *idealized* function, being the limit of a function of unit area defined in the range $|x-a| < \varepsilon$, as $\varepsilon \to 0$; outside this range, its value is zero. Its Laplace transform is e^{-pa}; it is useful in practice as yielding correct answers, though its rigorous basis and treatment cannot be examined in terms of elementary mathematics.

The step function. The symbol $H(x-a)$ denotes the value $H = 0$ for $0 < x < a$ and $H = 1$ for $x > a$. Its Laplace transform is e^{-pa}/p. Note the identity

$$\mathscr{L}[y(x-a)H(x-a)] = e^{-pa}\mathscr{L}[y(x)].$$

A function $y(x)$, normally defined for $x > 0$, may be replaced by zero in the range $0 < x < a$ by forming the product $y(x)H(x-a)$. Functions can be 'started' and 'stopped' by means of this device.

Periodic functions. If $y(x)$ is periodic with period 2π, then

$$\bar{y}(p) = \frac{1}{1-e^{-2\pi p}} \int_0^{2\pi} e^{-px} y(x)\, dx,$$

namely integration extends only from 0 to 2π instead of from 0 to ∞. If $Y(x)$ is defined to equal $y(x)$ for $0 < x < 2\pi$ and zero for $x > 2\pi$, then $\bar{y}(p) = \bar{Y}(p)/(1-e^{-2\pi p})$.

Problem 6.19 If $y(x)$ is a function of period 2π, consisting of a positive delta function at $x = \frac{1}{2}\pi$ and a negative delta function at $x = 3\pi/2$, show that its Laplace transform is $\frac{1}{2}\operatorname{sech}\frac{1}{2}\pi p$.

Solution. For $0 < x < 2\pi$, we have

$$y(x) = \delta(x - \tfrac{1}{2}\pi) - \delta(x - 3\pi/2),$$

and clearly

$$Y(x) = \delta(x - \tfrac{1}{2}\pi) - \delta(x - 3\pi/2).$$

Hence

$$\bar{y}(p) = \frac{1}{1 - e^{-2\pi p}}\overline{Y}(p) = \frac{e^{-\frac{1}{2}\pi p} - e^{-3\pi p/2}}{1 - e^{-2\pi p}}$$

$$= \frac{e^{-\frac{1}{2}\pi p}(1 - e^{-\pi p})}{(1 - e^{-\pi p})(1 + e^{-\pi p})} = \frac{1}{e^{\frac{1}{2}\pi p} + e^{-\frac{1}{2}\pi p}}$$

$$= \tfrac{1}{2}\operatorname{sech}\tfrac{1}{2}\pi p. \qquad \square$$

Problem 6.20 Find the Laplace transform of the periodic function of period a and given by $y = x$ $(0 < x < a)$.

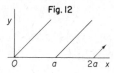

Fig. 12

Solution. Using the formula that employs the first period $0 < x < a$ only:

$$\bar{y}(p) = \frac{1}{1 - e^{-ap}}\int_0^a e^{-px}y(x)\,dx = \frac{1}{1 - e^{-ap}}\int_0^a xe^{-px}\,dx$$

$$= \frac{1}{1 - e^{-ap}}\left[-\frac{xe^{-px}}{p}\right]_0^a + \frac{1}{1 - e^{-ap}}\int_0^a \frac{e^{-px}\,dx}{p} \qquad \text{(by parts)}$$

$$= -\frac{ae^{-ap}}{p(1 - e^{-ap})} + \frac{1}{1 - e^{-ap}}\left[\frac{e^{-px}}{-p^2}\right]_0^a$$

$$= \frac{1}{p^2} - \frac{ae^{-\frac{1}{2}ap}\operatorname{cosech}\frac{1}{2}ap}{2p}. \qquad \square$$

Problem 6.21 Find the inverse Laplace transform of

$$\bar{y}(p) = e^{-2p}/(p^2 - 4p + 5).$$

Solution. The fact that the factor e^{-2p} occurs shows that a step function is involved, and that we must use the identity

$$\mathscr{L}[y(x - 2)\,H(x - 2)] = e^{-2p}\mathscr{L}[y(x)].$$

We first find the inverse of $1/(p^2 - 4p + 5)$, namely of $1/[(p - 2)^2 + 1]$; this is $e^{2x}\sin x$. The identity then yields

$$y(x) = e^{2(x - 2)}\sin(x - 2)\,H(x - 2),$$

a function 'starting' at $x = 2$. $\qquad \square$

Problem 6.22   Solve the equation
$$D^2y - 4Dy + 5y = 2\delta(x-2),$$
with $y(0) = Dy(0) = 1$.

Solution. The Laplace transform is
$$(p^2\bar{y} - 1 - p) - 4(p\bar{y} - 1) + 5\bar{y} = 2e^{-2p},$$
$$\bar{y} = \frac{p-3}{p^2 - 4p + 5} + \frac{2e^{-2p}}{p^2 - 4p + 5},$$
the former yielding the 'complementary function' and the latter the 'particular integral'. Rearranged, we have
$$\bar{y} = \frac{p-2}{(p-2)^2 + 1} - \frac{1}{(p-2)^2 + 1} + \frac{2e^{-2p}}{(p-2)^2 + 1}.$$
Taking the inverse transform, and quoting the result of the previous example, we conclude that
$$y(x) = e^{2x}\cos x - e^{2x}\sin x + 2H(x-2)\,e^{2(x-2)}\sin(x-2). \qquad \square$$

We see that the 'complementary function' continues indefinitely from $x = 0$ onwards. The 'particular integral' suddenly 'starts' at $x = 2$, preserving continuity of $y(x)$ at $x = 2$ but introducing a discontinuous derivative Dy at $x = 2$. This is characteristic of an impulse. In dynamics, an impulse acting on a particle does not instantaneously change the position of the particle, but its speed is suddenly changed at the moment of the impulse.

Problem 6.23   A constant e.m.f. E is introduced into a series L, R circuit between the times $t = a$ and $t = b$. Find the current in the circuit for $t > b$, assuming that it equals J initially.

Solution. Using $D \equiv d/dt$, we write the equation for the current j in the form
$$LD_j + R_j = 0 \quad (t < a), \qquad E \quad (a < t < b), \qquad 0 \quad (t > b).$$
Its Laplace transform is given by
$$(Lpj - J) + Rj = E(e^{-ap} - e^{-bp})/p,$$
where we have taken the e.m.f. to equal $EH(t-a) - EH(t-b)$. Hence
$$j = \frac{J}{Lp + R} + \frac{E(e^{-ap} - e^{-bp})}{p(Lp + R)}.$$
The inverse transform of the first term is $(J/L)e^{-Rt/L}$.

The inverse transform of
$$\frac{1}{p(Lp + R)} \equiv \frac{1}{R}\left(\frac{1}{p} - \frac{L}{Lp + R}\right)$$

80

is $1 - e^{-Rt/L}$; hence the inverse transform of

$$e^{-ap}\left(\frac{1}{p} - \frac{L}{Lp+R}\right)$$

is $H(t-a)[1 - e^{-R(t-a)/L}]$.

Totally, we conclude that

$$j(t) = \frac{Je^{-Rt/L}}{L} + \frac{EH(t-a)[1 - e^{-R(t-a)/L}]}{R} - \frac{EH(t-b)[1 - e^{-R(t-b)/L}]}{R}.$$

When $t > b$, the two H functions equal unity. Hence

$$j(t) = (J/L)e^{-Rt/L} + Ee^{-Rt/L}(e^{Rb/L} - e^{Ra/L})/R. \qquad \square$$

EXERCISES

1. Using the method of the Laplace transform, solve the following equations with the given initial conditions:

   (a) $D^2y - 3Dy + 2y = e^{3x}$, $y = Dy = \frac{1}{2}$ when $x = 0$.
   (b) $D^2y - 4y = 4e^{2x}$, $y = 4$, $Dy = -1$ when $x = 0$.
   (c) $D^2y - 2Dy + y = 12x^2e^x$, $y = 0$, $Dy = 1$ when $x = 0$.
   (d) $D^2y - 4Dy + 3y = 12\sin 3x - 6\cos 3x$, $y = 2$, $Dy = 3$
   when $x = 0$.

2. Solve the two simultaneous equations

$$3Du + 2Dv - u + v = e^x, \qquad 2Du + Dv - u + v = 5\sin x,$$

such that $u = v = 0$ when $x = 0$.

3. It is given that $u = 1$, $v = 2$ just before the unit impulse $\delta(t)$ is applied at $t = 0$ to the system governed by the equations

$$3Du + 4Dv + u - 12v = 2\delta(t), \qquad 2Du + 3Dv + u - 6v = 0.$$

Find u and v for $t > 0$.

Answers

1. (a) $y = \frac{1}{2}e^{3x} + e^x - e^{2x}$.
   (b) $y = xe^{2x} + 4\cosh 2x - \sinh 2x$.
   (c) $y = xe^x + x^4e^x$.
   (d) $y = e^{3x} + \cos 3x$.
2. $u = 2e^x + e^{2x} - 3 - 5\sin x$, $v = -e^x - e^{2x} - 3 + 5\sin x + 5\cos x$.
3. $u = 4e^{-2t} + 3e^{-3t}$, $v = -e^{-2t} - e^{-3t}$.

Chapter 7

Bessel Functions

Bessel's equation of order v,

$$D^2 y + x^{-1} Dy + (1 - v^2 x^{-2}) y = 0,$$

possesses two independent power-series solutions when v is not an integer:

$J_v(x)$ commences with the term $\dfrac{x^v}{2^v \Gamma(v+1)} \equiv \dfrac{x^v}{2^v(v)!}$;

$J_{-v}(x)$ commences with the term $\dfrac{x^{-v}}{2^{-v} \Gamma(-v+1)} \equiv \dfrac{x^{-v}}{2^{-v}(-v)!}$.

Here, $\Gamma(z)$ denotes the *gamma function*, defined as an integral when $\mathrm{Re}\, z > 0$:

$$\Gamma(z) = \int_0^\infty e^{-t} t^{z-1} \, dt.$$

Integration by parts shows that this function satisfies the recurrence relation $\Gamma(z) = (z-1)\Gamma(z-1)$, and since $\Gamma(1) = 1$, we see that

$$\Gamma(z) = (z-1)(z-2)\ldots 2 \cdot 1 \cdot \Gamma(1) = (z-1)!$$

when z is an integer. Even when z is an integer, many authors write $\Gamma(z) \equiv (z-1)!$ In particular, $\Gamma(\tfrac{1}{2}) = \sqrt{\pi}$; (let $t = u^2$ in the above integral).

A useful linear combination is the solution

$$Y_v(x) \equiv \cot v\pi \, J_v(x) - \operatorname{cosec} v\pi \, J_{-v}(x).$$

If v is an integer, n say, $J_{-n}(x)$ is proportional to $J_n(x)$, but $Y_v(x)$ tends to an independent solution $Y_n(x)$ involving $\log x$.

When $|x|$ is large, further linear combinations are useful. The Hankel functions are defined by

$$H_v^{(1)}(x) = J_v(x) + iY_v(x), \qquad H_v^{(2)}(x) = J_v(x) - iY_v(x).$$

The choice of solution in a particular problem depends both on the tables available and on the conditions imposed at infinity or at the origin. If $\mathrm{Re}\, v > 0$, $J_{-v}(x)$ and $Y_v(x)$ are rejected if the origin is included in the domain of the problem. If x is large (and positive, say), physical boundary conditions based on the following formulae will determine the type of solution used:

$$H_v^{(1)}(x) \doteqdot \sqrt{(2/\pi x)} \exp\left[i(x - \tfrac{1}{2}v\pi - \tfrac{1}{4}\pi)\right],$$
$$H_v^{(2)}(x) \doteqdot \sqrt{(2/\pi x)} \exp\left[-i(x - \tfrac{1}{2}v\pi - \tfrac{1}{4}\pi)\right],$$
$$J_v(x) \doteqdot \sqrt{(2/\pi x)} \cos(x - \tfrac{1}{2}v\pi - \tfrac{1}{4}\pi).$$

7.1 Differential Equations

Problem 7.1 Solve Bessel's equation of order $\frac{1}{2}$ by introducing the change of dependent variable $y = x^{-\frac{1}{2}}u$.

Solution. We substitute

$$y = x^{-\frac{1}{2}}u, \qquad Dy = x^{-\frac{1}{2}}Du - \tfrac{1}{2}x^{-\frac{3}{2}}u,$$
$$D^2y = x^{-\frac{1}{2}}D^2u - x^{-\frac{3}{2}}Du + \tfrac{3}{4}x^{-\frac{5}{2}}u$$

into the equation

$$D^2y + x^{-1}Dy + (1 - \tfrac{1}{4}x^{-2})y = 0,$$

obtaining after multiplying by $x^{\frac{1}{2}}$:

$$D^2u - x^{-1}Du + \tfrac{3}{4}x^{-2}u + x^{-1}(Du - \tfrac{1}{2}x^{-1}u) + (1 - \tfrac{1}{4}x^{-2})u = 0,$$

or
$$D^2u + u = 0.$$

Hence $u = \sin x$, $\cos x$, and $y_1 = x^{-\frac{1}{2}}\sin x$ and $y_2 = x^{-\frac{1}{2}}\cos x$.

The solution y_1 commences with the term $x^{\frac{1}{2}}$, so

$$J_{\frac{1}{2}}(x) = \frac{\sin x}{2^{\frac{1}{2}}\Gamma(\frac{3}{2})\sqrt{x}} = \left(\frac{2}{\pi x}\right)^{\frac{1}{2}}\sin x,$$

and y_2 commences with $x^{-\frac{1}{2}}$, so

$$J_{-\frac{1}{2}}(x) = \frac{\cos x}{2^{-\frac{1}{2}}\Gamma(\frac{1}{2})\sqrt{x}} = \left(\frac{2}{\pi x}\right)^{\frac{1}{2}}\cos x,$$

since $\Gamma(\frac{3}{2}) = \frac{1}{2}\sqrt{\pi}$, $\Gamma(\frac{1}{2}) = \sqrt{\pi}$. $\qquad\qquad\square$

Problem 7.2 Find the normal form for Bessel's equation.

Solution. Following the method given in section 2.4, we introduce into Bessel's equation the substitution

$$y = p(x)w, \qquad Dy = p\,Dw + (Dp)\,w,$$
$$D^2y = p\,D^2w + 2(Dp)\,Dw + (D^2p)\,w.$$

Then after division by p:

$$D^2w + \left(\frac{2Dp}{p} + \frac{1}{x}\right)Dy + \left(\frac{D^2p}{p} + \frac{Dp}{p} + 1 - \frac{v^2}{x^2}\right)y = 0.$$

We choose
$$\frac{2Dp}{p} + \frac{1}{x} = 0,$$

integrating to $\qquad 2\log p + \log x = \text{constant},$

or $p = x^{-\frac{1}{2}}$, say. It follows that $(Dp)/p = -\frac{1}{2}x^{-1}$ and $(D^2p)/p = \frac{3}{4}x^{-2}$, leading to the normal form

$$D^2w + [1 - (v^2 - \tfrac{1}{4})/x^2]w = 0.$$

83

Solutions of this equation are
$$w = yp^{-1} = x^{\frac{1}{2}}y = x^{\frac{1}{2}}J_\nu(x) \text{ and } x^{\frac{1}{2}}J_{-\nu}(x).$$ ∎

Problem 7.3 For large values of $|x|$, find approximate solutions of Bessel's equation.

Solution. The normal form may be rewritten approximately as
$$D^2w+w = 0,$$
neglecting x^{-2} for large $|x|$. Hence $w = \sin x, \cos x$, and
$$y = x^{-\frac{1}{2}}\sin x, x^{-\frac{1}{2}}\cos x.$$

These forms evidently are connected with the Bessel functions. Alternatively, $y = x^{-\frac{1}{2}}e^{\pm ix}$, connected with the Hankel functions.

Note. This method of procedure does *not* show how the approximate solutions are immediately related to the power-series forms for $J_\nu(x)$ and $J_{-\nu}(x)$. ∎

Problem 7.4 By extracting e^{ix}, say, from the function w in problem 7.2, find the descending-series (asymptotic) solutions for Bessel's equation.

Solution. Let $w = e^{ix}u$, so
$$Dw = e^{ix}Du + ie^{ix}u,$$
$$D^2w = e^{ix}D^2u + 2ie^{ix}Du - e^{ix}u.$$

Hence, substituting and dividing by e^{ix},
$$D^2u + 2iDu - \frac{\nu^2 - \frac{1}{4}}{x^2}u = 0.$$

Formally assume the descending development
$$u = \sum_{r=0}^{\infty} a_r/x^r,$$
yielding upon substitution
$$\sum \frac{r(r+1)a_r}{x^{r+2}} - 2i \sum \frac{ra_r}{x^{r+1}} - \frac{\nu^2 - \frac{1}{4}}{x^2} \sum \frac{a_r}{x^r} = 0.$$

Equating coefficients of $x^{-2}, x^{-3}, \ldots$, we obtain
$$2ia_1 = -[0.1 - (\nu^2 - \tfrac{1}{4})]a_0,$$
$$2ia_2 = -[1.2 - (\nu^2 - \tfrac{1}{4})]a_1,$$
$$2ia_3 = -[2.3 - (\nu^2 - \tfrac{1}{4})]a_2,$$

and so on. Solving for $a_1, a_2, \ldots$ in turn, and simplifying, we obtain the

84

coefficients,

$$a_1 = (-1)(\tfrac{1}{8}i)(4v^2-1)a_0,$$
$$a_2 = (-1)^2(\tfrac{1}{8}i)^2(4v^2-1)(4v^2-3^2)a_0,$$
$$a_3 = (-1)^3(\tfrac{1}{8}i)^3(4v^2-1)(4v^2-3^2)(4v^2-5^2)a_0.$$

This series (and the conjugate solution) are not convergent series, and they are known as *asymptotic series*. ◻

Problem 7.5   Express solutions of the Airy equation $D^2y = xy$ in terms of Bessel functions, the equation arising approximately when waves propagating in a plane-stratified inhomogeneous medium are reflected by a region in which the refractive index vanishes in one plane.

Solution. In elementary work, the changes of variable required are not obvious, and must therefore be supplied. First change the dependent variable to w where $y = x^{\frac{1}{2}}w$; then change the independent variable to t, where $t = \tfrac{2}{3}ix^{\frac{3}{2}}$.

First, we have

$$Dy = x^{\frac{1}{2}}Dw + \tfrac{1}{2}x^{-\frac{1}{2}}w, \qquad D^2y = x^{\frac{1}{2}}D^2w + x^{-\frac{1}{2}}Dw - \tfrac{1}{4}x^{-\frac{3}{2}}w;$$

hence, upon substitution,

$$D^2w + \frac{1}{x}Dw - \left(x + \frac{1}{4x^2}\right)w = 0.$$

Secondly, we have

$$\frac{dw}{dx} = \frac{dw}{dt}\frac{dt}{dx} = \frac{dw}{dt}ix^{\frac{1}{2}}, \qquad \frac{d^2w}{dx^2} = \frac{d^2w}{dt^2}(ix^{\frac{1}{2}})^2 + \frac{dw}{dt}(\tfrac{1}{2}ix^{-\frac{1}{2}}).$$

Substitution and division by $-x$ yield

$$\frac{d^2w}{dt^2} - \frac{3i}{2x^{\frac{3}{2}}}\frac{dw}{dt} + \left(1 + \frac{1}{4x^3}\right)w = 0,$$

or in terms of t,

$$\frac{d^2w}{dt^2} + \frac{1}{t}\frac{dw}{dt} + \left(1 - \frac{1}{9t^2}\right)w = 0.$$

This is Bessel's equation of order $\tfrac{1}{3}$, so

$$w = J_{\frac{1}{3}}(t), J_{-\frac{1}{3}}(t),$$
$$y = x^{\frac{1}{2}}J_{\frac{1}{3}}(\tfrac{2}{3}ix^{\frac{3}{2}}), x^{\frac{1}{2}}J_{-\frac{1}{3}}(\tfrac{2}{3}ix^{\frac{3}{2}}).$$

No singularity can occur at the origin; the first series commences with a term in x, and the second with a constant. ◻

Problem 7.6   In the propagation of waves in inhomogeneous plane-stratified media, the differential equation

$$D^2y + k^2(1 - e^{\alpha x})y = 0$$

arises; α is a positive constant. Find its solution in terms of Bessel functions.

Solution. We introduce the change of independent variable $t = ae^{bx}$, where a and b are to be found. Then

$$\frac{dy}{dx} = \frac{dy}{dt}\frac{dt}{dx} = \frac{dy}{dx}abe^{bx} = \frac{dy}{dt}bt, \qquad \frac{d^2y}{dx^2} = \frac{d^2y}{dt^2}(bt)^2 + \frac{dy}{dt}b(bt).$$

The differential equation becomes

$$\frac{d^2y}{dt^2}b^2t^2 + \frac{dy}{dt}b^2t + k^2\left[1 - \left(\frac{t}{a}\right)^{\alpha/b}\right]y = 0,$$

$$\frac{d^2y}{dt^2} + \frac{1}{t}\frac{dy}{dt} + \left[-\frac{k^2}{t^{2-\alpha/b}a^{\alpha/b}b^2} + \frac{k^2}{b^2t^2}\right]y = 0.$$

This equation may be reduced to Bessel's equation by the choice of parameters

$$2 - \alpha/b = 0, \qquad -k^2/a^{\alpha/b}b^2 = 1,$$

giving

$$b = \tfrac{1}{2}\alpha, \qquad a = 2ki/\alpha.$$

Thus the final equation is

$$\frac{d^2y}{dt^2} + \frac{1}{t}\frac{dy}{dt} + \left(1 + \frac{4k^2}{\alpha^2t^2}\right)y = 0,$$

with solutions

$$y = J_{2ki/\alpha}(t) \equiv J_{2ki/\alpha}\left[(2ki/\alpha)\,e^{\pm\frac{1}{2}\alpha x}\right]$$

together with a companion function. The nature of the physical problem determines which kinds of Bessel functions should be employed. $\quad\square$

Problem 7.7 Find the two independent power-series solutions of Bessel's equation when v is not an integer.

Solution. We substitute the power series

$$y = x^c \sum a_r x^r$$

into the differential equation, obtaining

$$\sum a_r(r+c)(r+c-1)x^{r+c-2} + \sum a_r(r+c)x^{r+c-2}$$
$$+ \sum a_r x^{r+c} - v^2 \sum a_r x^{r+c-2} = 0.$$

We equate to zero the coefficients of $x^{c-2}, x^{c-1}, \ldots$, obtaining

$$a_0c(c-1) \qquad +a_0c \qquad -v^2a_0 = 0,$$
$$a_1(c+1)c \qquad +a_1(c+1) \qquad -v^2a_1 = 0,$$
$$a_2(c+2)(c+1) + a_2(c+2) + a_0 - v^2a_2 = 0,$$
$$a_3(c+3)(c+2) + a_3(c+3) + a_1 - v^2a_3 = 0,$$

whence

$$a_0[c^2 - v^2] = 0, \qquad a_2[(c+2)^2 - v^2] = -a_0,$$
$$a_1[(c+1)^2 - v^2] = 0, \qquad a_3[(c+3)^2 - v^2] = -a_1, \qquad \text{etc.}$$

Hence $c = \pm v$ and $a_1 = 0$. It follows that $a_r = 0$ whenever r is odd. When r is even, $2m$ say,

$$a_{2m} = -\frac{a_{2m-2}}{[(c+2m)^2 - v^2]}$$

$$= (-1)^2 \frac{a_{2m-4}}{[(c+2m)^2 - v^2][(c+2m-2)^2 - v^2]}$$

$$= (-1)^m \frac{a_0}{[(c+2m)^2 - v^2][(c+2m-2)^2 - v^2]\dots[(c+2)^2 - v^2]}.$$

When $c = v$, we choose a_0 to equal $(\tfrac{1}{2})^v/\Gamma(v+1)$, yielding

$$a_{2m} = \frac{a_0}{2^{2m}m!\,(v+m)(v+m-1)\dots(v+1)} = \frac{(-1)^m(\tfrac{1}{2})^v}{2^{2m}m!\,\Gamma(v+m+1)};$$

hence

$$J_v(x) = (\tfrac{1}{2}x)^v \sum_{m=0}^{\infty} \frac{(-\tfrac{1}{4}x^2)^m}{m!\,\Gamma(v+m+1)}.$$

Similarly, when $c = -v$,

$$J_{-v}(x) = (\tfrac{1}{2}x)^{-v} \sum_{m=0}^{\infty} \frac{(-\tfrac{1}{4}x^2)^m}{m!\,\Gamma(-v+m+1)}.$$

These two series are convergent for all x, with the exception that one series is not defined at the origin. ☐

7.2 Applications

Problem 7.8   A circular plate is in uniform tension and clamped around its rim $r = a$. Investigate the normal modes of oscillation which are functions of r and time t only.

Solution. The governing partial differential wave equation in plane polar coordinates (with θ absent) is

$$\frac{\partial^2 z}{\partial r^2} + \frac{1}{r}\frac{\partial z}{\partial r} = \frac{1}{c^2}\frac{\partial^2 z}{\partial t^2},$$

where z denotes the displacement of a point of the plate normal to the equilibrium position.

The normal modes are found by considering a separated solution in the form

$$z = R(r)\,T(t),$$

the two functions R and T being functions of the one variable r and t

respectively, as indicated. Using $D_r \equiv d/dr$ and $D_t \equiv d/dt$, we obtain

$$(D_r^2 R)\, T + \frac{1}{r}(D_r R)\, T = \frac{1}{c^2} R D_t^2 T,$$

or, upon division by RT:

$$\frac{D_r^2 R + r^{-1} D_r R}{R} = \frac{1}{c^2}\frac{D_t^2 T}{T}.$$

Since the left-hand side is a function of r only, and the right-hand side is a function of t only, we conclude that both sides must equal a constant, $-n^2$ say. Thus

$$D_t^2 T + c^2 n^2 T = 0$$

with solutions $T = e^{\pm icnt}$ $(n \neq 0)$, and

$$D_r^2 + r^{-1} D_r R + n^2 R = 0.$$

If we write $nr = s$, we obtain Bessel's equation of order zero:

$$\frac{d^2 R}{ds^2} + \frac{1}{s}\frac{dR}{ds} + R = 0,$$

possessing a solution $R = J_0(s) = J_0(nr)$ together with its companion function $Y_0(nr)$. The general solution is

$$R = AJ_0(nr) + BY_0(nr).$$

But the origin $r = 0$ is included in the domain of the problem, so we must choose $B = 0$ to avoid the singularity at $r = 0$ occurring in $Y_0(nr)$.

Hence, choosing the time factor e^{inct}, we have the representation of a normal mode

$$z = AJ_0(nr)\, e^{inct}.$$

Finally, when $r = a$, z vanishes for all t, so

$$J_0(na) = 0,$$

yielding a series of values of na (extracted from tables of the zeros of Bessel functions). If λ_i denotes the ith zero for Bessel functions of order 0, then the $n_i = \lambda_i/a$ form a spectrum of values of n, corresponding to the frequencies $n_i c/2\pi$.

The actual oscillation of a circular plate will consist of a linear combination of such normal modes, together with further normal modes obtained by considering variations in the polar angle θ. $\qquad\square$

Problem 7.9 Waves propagated in a varying medium are governed by the normal second order differential equation

$$D^2 y + k^2(1 - e^{2\alpha x})y = 0.$$

The solution tends to zero as x tends to plus infinity. Examine the waves in the uniform medium when x is large and negative.

Solution. The governing differential equation is similar to that solved in example 7.6. Solutions are Bessel functions of order ki/α and argument

$$s \equiv (ki/\alpha)e^{\alpha x} \equiv (e^{\frac{1}{2}\pi i}k/\alpha)e^{\alpha x},$$

to be specific as pertaining to the argument of i, since the Bessel functions are not single-valued; it is essential to define a branch (arbitrarily) and to remain on it throughout the investigation of the problem.

The formulae quoted at the beginning of the chapter contain all the information required to solve the problem. There, the asymptotic form of $H_\nu^{(1)}(x)$ contains the factor e^{ix}. In the present case, $H_{ki/\alpha}^{(1)}[(e^{\frac{1}{2}\pi i}k/\alpha)e^{\alpha x}]$ contains the factor $\exp[-(k/\alpha)e^{\alpha x}]$, which tends to zero as $x \to +\infty$. But the Hankel function of the second kind tends to infinity, and is therefore rejected from the general solution. Thus we take

$$
\begin{aligned}
y \propto H_\nu^{(1)}(s) \ &= \ J_\nu(s) + i Y_\nu(s) \\
&= \ J_\nu(s)\,(1 + i\cot\nu\pi) - \operatorname{cosec}\nu\pi J_{-\nu}(s) \\
&\propto \ e^{-i\nu\pi} J_\nu(s) - J_{-\nu}(s).
\end{aligned}
$$

As $x \to -\infty$, $s \to 0$, and so all terms in the power series for $J_\nu(s)$ and $J_{-\nu}(s)$ vanish, except the first terms in each case. Hence

$$
\begin{aligned}
y &\to e^{-i\nu\pi}\,\frac{(\tfrac{1}{2}s)^\nu}{\Gamma(\nu+1)} - \frac{(\tfrac{1}{2}s)^{-\nu}}{\Gamma(-\nu+1)} \\
&= \frac{e^{-i\nu\pi}}{\Gamma(\nu+1)}\left(\frac{e^{\frac{1}{2}\pi i}k}{2\alpha}\right)^\nu e^{\nu\alpha x} - \frac{1}{\Gamma(-\nu+1)}\left(\frac{e^{\frac{1}{2}\pi i}k}{2\alpha}\right)^{-\nu} e^{-\nu\alpha x}.
\end{aligned}
$$

But $e^{\nu\alpha x} = e^{ikx}$, representing one wave present, and $e^{-\nu\alpha x} = e^{-ikx}$, representing the other wave. The ratio of the amplitudes of these two waves is often required in physical problems; its value is given by

$$
R = -\frac{e^{-i\nu\pi}}{\Gamma(\nu+1)}\left(\frac{e^{\frac{1}{2}\pi i}k}{2\alpha}\right)^\nu \Big/ \frac{1}{\Gamma(-\nu+1)}\left(\frac{e^{\frac{1}{2}\pi i}k}{2a}\right)^{-\nu} = \frac{\Gamma(-ik/\alpha)}{\Gamma(ik/\alpha)}\left(\frac{k}{2\alpha}\right)^{2ik/\alpha}
$$

upon simplification. ◻

Problem 7.10   A heavy flexible chain of length a is suspended under gravity from a fixed point A. Find the normal modes of oscillation when the chain executes small transverse oscillations in a vertical plane.

Solution. Let O denote the lower end of the chain in the static equilibrium position; x is measured vertically from O, and y denotes the horizontal displacement in the x, y-plane. If m is the line density of mass, the horizontal rate of change of momentum of an element PQ is $m\,\delta x\,\partial^2 y/\partial t^2$.

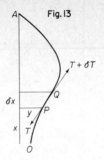

Fig. 13

At a point P, the horizontal component of the tension T is

$$T \sin \psi \doteq T \tan \psi = T \, \partial y / \partial x$$

for small displacements. At Q, the horizontal component is

$$T \frac{\partial y}{\partial x} + \frac{\partial}{\partial x} \left(T \frac{\partial y}{\partial x} \right) \delta x,$$

so the total component of force is $(\partial/\partial x)(T \, \partial y/\partial x)\delta x$. But $T = mgx$ (the weight of the portion OP), so the equation of motion is

$$\frac{\partial}{\partial x} \left(mgx \frac{\partial y}{\partial x} \right) = m \frac{\partial^2 y}{\partial x^2},$$

or

$$x \frac{\partial^2 y}{\partial x^2} + \frac{\partial y}{\partial x} = \frac{1}{g} \frac{\partial^2 y}{\partial t^2}.$$

For harmonic time variations, let $y = X(x)\sin \omega t$; (this is a separated solution, and represents a normal mode). Then

$$x \, D^2 X + DX = -(\omega^2/g)X.$$

This is not Bessel's equation, but this may be recovered by the change of independent variable $z = 2\omega g^{-\frac{1}{2}} x^{\frac{1}{2}}$. Then

$$\frac{dX}{dx} = \frac{dX}{dz} \frac{dz}{dx} = \frac{dX}{dz} \omega g^{-\frac{1}{2}} x^{-\frac{1}{2}},$$

$$\frac{d^2 X}{dx^2} = \frac{d^2 X}{dz^2} (\omega g^{-\frac{1}{2}} x^{-\frac{1}{2}})^2 - \frac{dX}{dz} \tfrac{1}{2} \omega g^{-\frac{1}{2}} x^{-\frac{3}{2}},$$

yielding upon substitution,

$$\frac{d^2 X}{dz^2} + \frac{1}{z} \frac{dX}{dz} + X = 0.$$

This is Bessel's equation of order zero, with general solution

$$X = A J_0(z) + B Y_0(z).$$

90

When $x = 0$ (that is, $z = 0$), X is finite, so $B = 0$ to avoid the singularity caused by $Y_0(z)$. Hence

$$y = AJ_0(2\omega g^{-\frac{1}{2}}x^{\frac{1}{2}})\sin \omega t.$$

When $x = a$ at the upper end A, this end is fixed, so $y = 0$ for all time t. Thus

$$0 = J_0(2\omega g^{-\frac{1}{2}}a^{\frac{1}{2}}),$$

or

$$2\omega g^{-\frac{1}{2}}a^{\frac{1}{2}} = \lambda_i,$$

where λ_i denotes the ith zero of J_0 (obtainable from tables of the zeros of Bessel functions). Hence

$$\omega_i = \tfrac{1}{2}\lambda_i\sqrt{(g/a)},$$

yielding a spectrum of possible frequencies. The normal modes are finally proportional to

$$y = J_0[\lambda_i\sqrt{(x/a)}]\sin[\tfrac{1}{2}\lambda_i\sqrt{(g/a)}t].$$

The general solution consists of an arbitrary linear combination of the normal modes (including the corresponding forms with cosine replacing sine). $\qquad\qquad\qquad\qquad\qquad\qquad\qquad\qquad\square$

EXERCISES

1. Show that the equation $D^2y = x^2y$ is satisfied by

$$y = x^{\frac{1}{2}}J_{\frac{1}{4}}(\tfrac{1}{2}ix^2).$$

2. By changing the variables (as indicated in the answer) show that the general solution of the equation

$$x^2D^2y+(1-2a)xDy+[b^2c^2x^{2c}+(a^2-n^2c^2)]y = 0$$

is given by $y = Ax^aJ_n(bx^c)+Bx^aY_n(bx^c)$.

3. In a problem on the stability of a tapered strut, the displacement y satisfies the equation

$$D^2y+(k^2/4x)y = 0,$$

where k is to be found. If $y = x^{\frac{1}{2}}w$, $z = kx^{\frac{1}{2}}$, reduce this to Bessel's equation of order 1. If $Dx = 0$ when $x = a$ and b, show that the equation for k is

$$J_0(k\sqrt{a})Y_0(k\sqrt{b}) = J_0(k\sqrt{b})Y_0(k\sqrt{a}).$$

Assume that $D[zJ_1(z)] = zJ_0(z)$.

Index